Earth Science Week 2018: Earth as Inspiration

October 14-20, 2018
Highlights Report

American Geosciences Institute
4220 King Street
Alexandria, VA 22302 U.S.A.
www.americangeosciences.org
703-379-2480

If you have comments concerning this report, please contact:
Ed Robeck, Ph.D.
Director, Department of Education and Policy
Director, Center for Geoscience and Society
American Geosciences Institute
703-379-2480 x245
ecrobeck@americangeosciences.org

This report is designed to give an overview of the activities organized by AGI and other groups for Earth Science Week. We hope this information on 2018 events and publicity inspires you to develop your own activities next year. Please visit **www.earthsciweek.org** for event planning, materials, resources, and support. Contact Earth Science Week staff at **info@earthsciweek.org** for assistance in planning for Earth Science Week.

Table of Contents

Highlights Report: Earth Science Week 2018

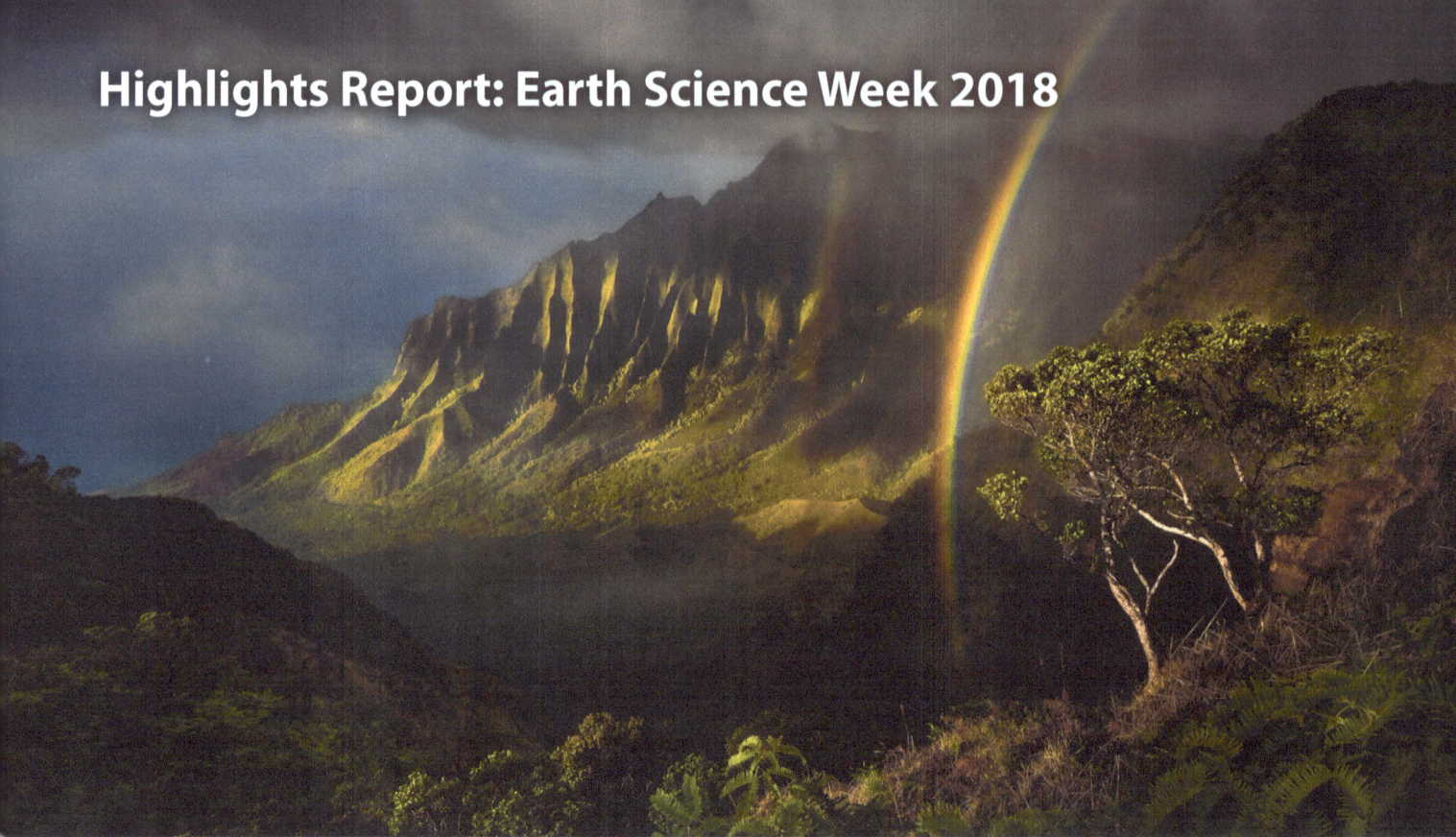

Earth Science Week 2018 Photo Contest entry by winner Matt Meisenheimer.

Introduction

Held October 14–20, 2018, the 21st annual **Earth Science Week** celebrated the theme of **"Earth as Inspiration."** The 2018 event promoted awareness of what geoscience tells us about human interaction with the planet's natural systems and processes. Learning resources and activities engaged young people and others in exploring the important relationships that exist between art and the geosphere (earth), hydrosphere (water), atmosphere (air), and biosphere (life). The 2018 theme promoted the inclusion of the arts into STEM learning and vice versa as a way to enhance Earth science learning through different avenues like music, sculpture, and poetry.

AGI organizes Earth Science Week as a service to member societies, with generous help from partners that provide funding, donate materials, organize events, and publicize the event. Funding partners in 2018 included the U.S. Geological Survey (USGS); National Aeronautics and Space Administration (NASA), National Park Service (NPS); American Association of Petroleum Geologists (AAPG) Foundation; American Geophysical Union (AGU); Society for Mining, Metallurgy, and Exploration (SME); Howard Hughes Medical Institute (HHMI); Association of American State Geologists (AASG); Geological Society of America (GSA); Archaeological Institute of America (AIA); AmericaView; Equinor; Minerals Education Coalition; Schlumberger; Water Footprint Calculator (Grace Communications Foundation); Consumer Energy Education Foundation; Keystone Policy Center; and TGS.

Earth Science Week **participation remained strong.** People in all 50 states and more than 20 countries participated in events and activities. The Earth Science Week website received over 550,000 page views in 2018. In addition, hundreds of people nationwide actively participated in the program's visual arts, video, essay, and photography contests.

Earth Science Week events ranged from educators teaching Earth science activities in their classrooms to public outreach events with local artists. A detailed list of events can be found in the second half of this report. This list represents only events reported directly to AGI, so please notify Earth Science Week staff if your participation is not listed.

Additional events are highlighted on the Earth Science Week website (www.earthsciweek.org/highlights), which features press releases and other items by members of the geoscience community, as well as news media promoting Earth Science Week. Television and radio news programs covered Earth Science Week on local stations in several states. Each year, web, print, and broadcast media coverage of Earth Science Week, along with direct outreach by AGI, reach more than **50 million people.**

Summary of Activities

Earth Science Week 2018 Photo Contest entry by Kate Ragusa.

Key Partnerships and Efforts

Earth Science Week's success depends on the collaboration of key partners. In 2018, AGI pursued signature initiatives and forged partnerships with several organizations (listed alphabetically):

Educators seeking teaching resources and other support were directed by AGI to the **American Association of Petroleum Geologists** (AAPG) and the **AAPG Foundation**, both longtime supporters of Earth Science Week. In addition, Earth Science Week promoted awareness of AAPG's Distinguished Lecturer and Teacher of the Year programs. AAPG Student Chapters received kits. Program participants were encouraged to read and use "Visiting Geoscientists: An Outreach Guide for Geoscience Professionals," a handbook co-produced by AGI and AAPG's Youth Education Activities Committee. Program participants were encouraged to attend AAPG's Annual Convention and Exhibition and compete to win the AAPG Teacher of the Year award. AAPG's "Investigating Rock Types" activity was featured in the Earth Science Week 2018 activity calendar.

The **American Geophysical Union** (AGU) continued its role as a supporting program partner in 2018 with the contribution of funds as well as expertise. Earth Science Week's 2018 activity calendar featured AGU's "Digging Into Soil" classroom activity. AGU's annual meetings, professional development workshops, programs for college students, print and electronic resources, GIFT workshops, AGU On-Demand, and "LEARN With AGU" video series were promoted through the Earth Science Week e-newsletter, website, and activity calendar. Earth Science Week materials were also given out at AGU's Fall Meeting in Washington, DC.

Earth Science Week 2018 promoted awareness of the **American Institute of Professional Geologists** (AIPG), an AGI member society that advocates for geologists and certifies their credentials. AIPG offers several PowerPoint presentations online for free download, presenting career information for young, newly graduated geoscientists. AIPG also provided a geologic-timescale bookmark for the educator kit.

Program participants were encouraged to go online to conduct an activity called "Climate at a Glance: From Local to National Scale" that the **American Meteorological Society** (AMS) created in cooperation with NOAA's National Centers for Environmental Information. The activity introduces students to the NOAA Climate at a Glance website, which allows real-time analysis of monthly temperature and precipitation data nationwide.

The **Archaeological Institute of America** (AIA), a continuing Earth Science Week partner and supporter, provided

a classroom activity on "Earth and Ancient Architecture" for the program's 2018 activity calendar. In addition, the program promoted awareness of and participation in AIA's International Archaeology Day, which takes place annually on the final day of Earth Science Week.

Earth Science Week directed participants' attention to the **Association of American Geographers** (AAG), an AGI member society that offers an array of web resources for K–12 and college-level instruction. Materials support geographic approaches to Earth science education. For example, Geographic Advantage, an educational companion for the National Research Council's "Understanding the Changing Planet," shows students how geographers use maps to explore environmental change. AAG provided a fact sheet on its GeoMentor program and a postcard on its annual meeting for this year's educator kit.

The **Association of American State Geologists** partnered with AGI and the USGS to support Geologic Map Day during Earth Science Week 2018. State geologists nationwide made geologic maps of their states available on their websites for students to use in classroom activities on Geologic Map Day.

Earth Science Week promoted awareness of a website of great value to educators, AGI's **Center for Geoscience & Society.** Largely through its Education GeoSource (formerly Education Resources Network) database, the center enhances geoscience awareness across all sectors of society by generating new approaches to building geoscience knowledge, engaging the widest possible range of stakeholders, and creatively promoting existing and new resources and programs.

Program participants learned about three online videos by the **Center for Ocean Sciences Education Excellence** (COSEE) depicting dramatic changes in Alaska's marine ecosystems through interviews with scientists. The videos were produced by COSEE Alaska in cooperation with other geoscience organizations.

As promoted by Earth Science Week, the **Climate Literacy and Energy Awareness Network** (CLEAN) online portal stewards a major collection of climate and energy science educational resources and supports a community of professionals committed to improving climate and energy literacy. Key components include the CLEAN collection of climate and energy science resources, CLEAN guidance in teaching climate and energy science, and the CLEAN network of professionals committed to improving climate and energy literacy. For this year's Earth Science Week Toolkit, CLEAN provided a bookmark featuring a link to key resources.

Earth Science Week staff traveled to Houston to exhibit at the Energy Day festival hosted by program partner **Consumer Energy Alliance**, an advocacy organization that provides consumers with unbiased information on energy issues. AGI staff shared geoscience-based energy information with thousands of students, teachers, and other community members.

The **Critical Zone Observatories** provided an informational flyer on this National Science Foundation program, including NGSS-aligned classroom activities "I notice…" and "I wonder…" for the Earth Science Week Toolkit in 2018.

Earth Science Week participants were encouraged to celebrate **Earth Day** in April 2018 with classroom activities, experiments, and investigations exploring the science behind how the world works. Because Earth Science Week offers education materials, information, and tools throughout the year, school audiences were urged to make use of tools highlighting the theme of "Earth as Inspiration." In addition, program participants were invited to take part in a free live online event, hosted by the California Academy of Sciences in April, during which teachers and students were able to ask questions of "Academy Sustainability Scientists" studying these topics.

Science teachers were invited to take part in the fourth annual **Earth Educators' Rendezvous** in July 2018 at the University of Kansas, Lawrence, Kansas. The event's combination of workshops, posters, talks, round-table discussions, and plenary presentations helped guide participants through a suite of interrelated challenges characteristic of Earth education in schools, colleges, and universities today.

Former U.S. Secretary of the Interior Sally Jewell served as AGI's **Earth Science Education Ambassador** in 2018. In this role Jewell advocated for strong geoscience instruction in K-12 education. Jewell attended the Intel International Science and Engineering Fair (Intel ISEF), interacting with students as they presented research displayed at the fair in May. Shortly after that, Jewell attended AGI's and ExxonMobil's annual Teacher Leadership Academy for K-8 educators, providing instructional resources and inquiry-based hands-on activities. Finally, she rounded out the year by leading a collaborative field experience at Valles Caldera National Preserve, alongside The Geological Society of America (GSA), the New Mexico Bureau of Geology and Mineral Resources, the National Park Service (NPS), and eighth-grade students from the local Jemez Pueblo. The field trip introduced students in the area to the geoscience of a natural wonder that many of them had never seen, or understood from a scientific perspective, even though it was virtually in their back yard. These events are featured in a brand-new film series, Earth as Inspiration, in which Jewell discusses the many ways our Earth inspires us. These short films are now available for free viewing online.

For this year's Earth Science Week Toolkit, **EarthScope** provided an informational postcard on the personal stories of Earth scientists.

The **Earth Science Women's Network (ESWN)** hosted a Science-A-Thon in conjunction with Earth Science Week. This social media event showcases the many different people and activities involved in science. The goal is to increase visibility of scientists and the important work they do, showing what a "day of science" looks like, with participants posting photos throughout the day. The event was also used as a fundraiser for the organization.

ExxonMobil, a longtime Earth Science Week partner, continued its support of the program. During summer 2018, ExxonMobil Exploration and AGI partnered to hold two five-day Earth Science/STEM Teacher Leadership Academies in Houston. Each academy — one for K–5 teachers and one for middle-school teachers — provided educators with Earth science content, hands-on activities, resources and field experiences for them to use with their students in the classroom and with their colleagues in professional development settings.

Through Earth Science Week, participants learned about educational resources and programs of the **Geological Society of America** (GSA), a longtime program partner. Featured education and outreach programs included the GeoTeacher program, Teacher GeoVenture trips, the Distinguished Earth Science Teacher in Residence program, and GSA's GeoCorps America program. GSA also organized International EarthCache Day at the start of Earth Science Week 2018 and contributed as an active partner in the Geologic Map Day initiative. Earth Science Week's 2018 activity calendar included GSA's "Latitude and Longitude" activity. The educator kit also featured more GSA activities, "The Science and Technology of Gold."

Earth Science Week continued promoting **The Geological Society of London** (GSL), an AGI member society, map-based resources. GSL's map-based resources focus on plate tectonics and the rock cycle through accompanying webpages.

Program participants were directed to **Geology.com,** a major Earth Science Week partner which provides a variety of geoscience materials including daily Earth

Earth Science Week 2018 Visual Arts Contest entry by finalist Saachi Tamboli.

science news, maps, an online dictionary of Earth science terms, and information on geoscience careers, as well as resources for teachers, including links to lesson plans from major Earth science organizations. Geology.com, in turn, covered Earth Science Week announcements, programs, and activities throughout the year.

Google, an ongoing Earth Science Week partner, contributed a "Making Earth Art With Google Earth" activity to the Earth Science Week 2018 activity calendar.

The **Howard Hughes Medical Institute**, a continuing program partner, provided key materials for the Earth Science Week 2018 Toolkit. A prized component of this year's kit included a poster on "Understanding Global Change."

To help teachers and students delve into the science behind current events, Earth Science Week continued to direct them to the **Incorporated Research Institutions for Seismology** (IRIS) website. IRIS provides educational resources including PowerPoint presentations, animations, and visualizations, as well as links to Spanish-language materials and USGS data that deal with current events such as earthquakes. For the 2018 educator kit, IRIS also provided an activity sheet "Determining Earth's Layered Interior."

As program participants learned, ethical behaviors and practices are of vital concern to geoscientists. The **International Association for Promoting Geoethics** (IAPG), an international associate of AGI, celebrated the second annual International Geoethics Day on the Thursday of Earth Science Week 2018. To promote awareness of geoethics, IAPG offered documents such as The Geoethical Promise and The Cape Town Statement on Geoethics. IAPG also provided online resources such as Teaching GeoEthics Across the Geoscience Curriculum and various training programs.

The **Minerals Education Coalition** (MEC) of the **Society for Mining, Metallurgy, and Exploration,** an AGI member society, supported Earth Science Week in 2018. MEC provided 2018 calendar activity "Determining Mineral Reserves." And program participants received information about a series of MEC podcasts with industry experts, showing students how an interest in STEM subjects can lead to a rewarding career in the mining industry.

Earth Science Week directed participants' attention to a joint effort by AGI and the **National Association of Geoscience Teachers** (NAGT) to strengthen implementation of the *Next Generation Science Standards* (NGSS) at the state level. Science educators have regularly participated in free webinars and joined discussions online since the 2015 NGSS Summit. Earth Science Week participants have been invited to take advantage of NAGT offerings including online lessons, NAGT's Outstanding Earth Science Teacher Awards, the Dorothy Stout Professional Development Grants, and the Journal of Geoscience Education.

NASA, a founding partner of Earth Science Week, provided materials in the Earth Science Week 2018 Toolkit (and online) designed to frame phenomena-based student investigations. NASA included two resources in the kit, A Globe Program Cloud Identification Chart, as well as an activity, "Numbers to Pictures: How Satellite Images are Created." Earth Science Week's 2018 activity calendar also featured a "What Covers Our Land" classroom activity from NASA. Throughout the year, Earth Science Week promoted awareness of NASA's online offerings, such as NASA Wavelength, SciJinks, Space Place, and Climate Kids.

AGI staff at the 2018 National Fossil Day event on the National Mall in Washington, D.C.
Image credit: AGI/Joe Lilek

The **National Earth Science Teachers Association** (NESTA), a longtime Earth Science Week partner, continued its vital role in helping AGI promote excellence in geoscience education. At the National Science Teachers Association Annual Conference, the NESTA Reception included a ceremony during which a teacher was given the Edward C. Roy Jr. Award for Excellence in K–8 Earth Science Teaching. NESTA members also received copies of the Earth Science Week 2018 poster in their association newsletter. Finally, NESTA assembled an incredible Earth Science Week Collection of teaching resources for educators. Day by day, the collection offered instructional resources tailored for each of Earth Science Week's Focus Days, from Sunday's International EarthCache Day through Saturday's International Archaeology Day.

Earth Science Week raised awareness of **National Environmental Education Week** (EE Week), the nation's largest environmental education event. Focusing largely on STEM topics, EE Week connected educators with resources to promote K–12 students' understanding of the environment.

Earth Science Week promoted the **National Groundwater Association**'s (NGWA) Groundwater Awareness Week in March 2018 as well as NGWA's Protect Your Groundwater Day program in September 2018, advocating water conservation and safety. The AGI member society offers Groundwater Adventures, a website providing educational activities for young people.

Earth Science Week's 2018 activity calendar featured a "Citizen Science" activity from the **National Oceanic and Atmospheric Administration** (NOAA). In addition, Earth Science Week 2018 directed participants to NOAA's online multimedia education resources, including lesson plans, real-world data, instructional games, videos, and more.

A longtime Earth Science Week partner, the **National Park Service** (NPS) continued for the ninth year a major component to its involvement in Earth Science Week. National Fossil Day, established as a celebration to take place annually on the Wednesday of Earth Science Week, once again reached millions of people. Earth

Science Week promoted awareness of NPS's interactive Web Ranger program, which helps people of all ages learn about the national parks. Program participants also learned about the NPS National Natural Landmarks program, which recognizes and encourages the conservation of sites that contain outstanding biological and geological resources. NPS videos on climate change were made available to program participants. Posters illuminating the geologic resources relating to caves, geology of national parks, and plants benefiting from clean air, appearing in the Earth Science Week 2018 Toolkit, successfully continued the series of park posters produced collaboratively by the federal agency and AGI. Earth Science Week's 2018 activity calendar also featured a classroom activity on "Awesome Fossils."

Earth Science Week participants learned about K–12 soil resources offered online by the **Natural Resource Conservation Service** (NRCS). Resources for the elementary level include frequently asked questions even soil songs. Middle and high school level resources include soil facts, health and quality, and state-specific soil information. Lesson plans and other resources are also accompanied by links to other websites.

Earth Science Week partnered with the **National Science Teachers Association** (NSTA) again in 2018, reaching science educators nationwide. Program participants learned about "Freebies for Science Teachers" on the NSTA website. AGI also attended the NSTA Annual National Conference in Atlanta.

The Nature Conservancy, which offers informational resources ideal for educators aiming to teach about a wide range of geoscience topics, was promoted through Earth Science Week. Videos, webcasts, articles, and other materials conveyed the work of scientists engaged in conservation efforts around the world.

Gearing up for National Fossil Day, Earth Science Week directed program partners' attention to the **Paleontological Research Institution** (PRI), an AGI member society providing education materials and opportunities for science teachers and students at all grade levels. The online Teacher Friendly Guide, for example, gives brief geologic histories of every region of the United States.

For National Fossil Day, AGI collaborated with NPS partners and other geoscience organizations to conduct a National Fossil Day event in Washington, D.C. AGI educated and entertained visitors with a green screen-equipped "Paleontology Play Space" photo booth, funded by the **Paleontological Society,** on the National Mall.

AGI staff attended the 2018 National Science Teachers Association (NSTA) National Conference.
Image credit: AGI/ Sequoyah McGee

Longtime program partner **Partners in Resource Education** (PRE) provided activities focusing on the geoscience of conservation. The consortium of seven federal agencies educates thousands of young people, introduces them to natural resource careers, and cultivates the next generation of land and water stewards. In 2018, PRE collaborated to promote awareness of Earth Science Week, and vice versa.

Schlumberger Excellence in Educational Development (SEED) is a nonprofit education program that empowers educators to share their passion for learning and science with students. In addition to promoting awareness of SEED and other resources, AGI has partnered with the program to provide geoscience education resources in both Spanish and English since 2010.

For teachers aiming to "shake up" education, Earth Science Week shone a spotlight on the **Seismological Society of America** (SSA). SSA's website provided seismic eruption models, wave animations, plate tectonics simulations, information on tsunamis, and more. SSA also offered publications, information on seismology careers, a distinguished lecturer series, and an electronic encyclopedia of earthquakes.

Program participants received information about **Smithsonian Education,** which offers a fascinating exploration of Earth's soil with its "Dig It! The Secrets of Soil" exhibition. Information, videos, expert instruction, and activity sheets are available online.

Earth Science Week participants were encouraged to take advantage of offerings of the **Society of Exploration Geophysicists** (SEG), which provides programs for educators and students. For example, the distinguished lecturer series and honorary lecturer series both enabled students to meet professional geophysicists, learn about groundbreaking research in the field of seismic research, and obtain valuable career information.

Advanced by the **Society of Petroleum Engineers** (SPE), the Energy4Me program offers teachers a collection of tools for teaching about oil, gas, and other energy

AGI staged its first-ever exhibit of photos at Washington, D.C.'s Union Station.

Image credit: AGI/Sequoyah McGee

sources, including classroom activities, experiments, and presentations, as well as teacher workshops and energy education materials for the classroom.

The **Soil Science Society of America** (SSSA), a longtime program partner, provides lessons, activities, fun facts, sites of interest, and soil definitions for the novice soil scientist online. These resources were promoted by the October event. Earth Science Week's 2018 activity calendar featured a "Painting With Soil" classroom activity courtesy of SSSA.

The Earth Science Week program ended the year by announcing the next year's program theme, "**Geoscience Is For Everyone**," with great fanfare at **AGU**'s Fall Meeting in Washington, DC.

STEMIE (Science, Technology, Engineering, and Math linked to Invention and Entrepreneurship), an education framework that elevates youth invention and entrepreneurship education to a core part of K–12 education, maps essential unstructured problem-solving teaching activities to core STEM curricula and standards. AGI staff participated in judging geoscience entries in STEMIE's annual student competition.

SWITCH Energy Alliance provided an informational postcard on this video-based energy education project, including online resources, for the Earth Science Week 2018 Toolkit.

For the Earth Science Week 2018 Toolkit, **UNAVCO** provided a poster titled "What is Geodesy," including detailed information on the applications of geodesy in different industries.

The **U.S. Bureau of Land Management** (BLM), a continuing Earth Science Week partner, provided a flyer including a dinosaur activity sheet for the 2018 educator kit. BLM also was the subject of Earth Science Week promotions, including its Classroom Investigation Series online.

The **U.S. Department of Energy**'s Office of Energy Efficiency & Renewable Energy website offered classroom activities and materials for K–12 science instruction, as program participants learned. Additionally, educators were invited to explore DOE's websites for the National Renewable Energy Laboratory. Earth Science Week participants were made aware that AGI's Center for Geoscience & Society produced education materials, including videos in English and Spanish, education guides, a "quick start" guide to energy literacy, lesson connections, and guidance on aligning energy literacy lessons with the *Next Generation Science Standards*. Essential Principles and Fundamental Concepts for Energy Education resources were made available on the DOE website. Teams of program participants were urged to enter DOE's new Geothermal Design Challenge, which invites young people to design an infographic that illustrates how geothermal energy is clean, safe, reliable, and renewable.

Earth Science Week also promoted awareness of the **U.S. Environmental Protection Agency**'s collection of free resources to enhance middle school students' understanding of climate change impacts on the United States' wildlife and ecosystems. The online toolkit includes case studies, activities, and videos based on climate science, environmental education, and stewardship information.

Overlapping Earth Science Week 2018, National Wildlife Refuge Week was held October 14–20. The event, celebrating the richness of the 550 units that make up America's National Wildlife Refuge System, was sponsored once again by the **U.S. Fish and Wildlife Service** (FWS), an Earth Science Week partner.

Earth Science Week participants learned about online education resources offered by the **U.S. Geological Survey** (USGS), a longtime Earth Science Week partner and supporter, as well as the thousands of free images and over 69,000 searchable publications such as maps, books, and charts provided online by the agency. Earth Science Week's 2018 activity calendar featured a "ShakeAlert Earthquake Early Warning" classroom activity courtesy of USGS. Also, USGS continued its leadership role as a founding partner of Geologic Map Day in 2018, providing support as well as its National Geologic Map Database's MapView, which offers a mosaic view of published geologic maps.

Earth Science Week partnered with **Washington Union Station**, **Mazza Gallerie,** as well as Terminal B of the **George Bush Intercontinental Airport** in Houston to exhibit over 20 top photos entered in the program's 2017 photo contest, which explored the theme of "Earth and Human Activity Here." Located in the heart of the nation's capital, Union Station is one of the country's busiest train stations, with an average of about 100,000 train and local metro passengers passing through daily. Mazza Gallerie is a premier shopping destination in Washington, DC's Friendship Heights neighborhood. Located just outside of downtown Houston, the George Bush Intercontinental Airport serves over 110,000 passengers daily

Education resources of the **Water Environment Federation** (WEF) were promoted among Earth Science Week participants, especially during World Water Day in March. Program participants learned about clean water and real-life professionals who keep water resources safe. WEF is a nonprofit association that provides technical education and training for water quality professionals.

Earth Science Week participants in Canada were invited to take part in the 2018 **WHERE Challenge** sponsored by Teck Resources Limited. WHERE stands for the places where Earth scientists work: Water, Hazards, Energy, Resources, and Environment. The challenge, a national contest endorsed by the Canadian Earth sciences community, asked students ages 9–14: "What on Earth is in your stuff, and WHERE on Earth does it come from?" Students explored the nonrenewable Earth resources that make up their favorite objects.

Earth Science Week continued promoting awareness of **Windows on Earth,** an online educational project that features photographs taken by astronauts on the International Space Station. The site is operated by TERC, an educational non-profit, in collaboration with the Association of Space Explorers (the professional association of flown astronauts and cosmonauts), the Virtual High School, and CASIS (Center for Advancement of Science in Space). The images help show Earth from a global perspective.

Earth Science Week promoted PLAN!T NOW's **Young Meteorologist Program** taking students on a severe weather preparedness adventure. The program was developed in partnership with NOAA's National Weather Service and the National Education Association. Young Meteorologists were given opportunities to put their knowledge to work in hands-on activities and community service projects, learning about severe weather science and safety.

International Highlights

The first **Earth Science Week Japan** event was held in Shizuoka, with almost 2,000 people in attendance. Organizations such as the Japan Geoscience Union (JpGU), Shizuoka University, Shizuoka Earth Science Association, Shizuoka City, and others worked in cooperation with the Committee for Earth Science Week Japan to hold activities for the general public at various museums and outside locations. Activities included special talks with local scientists, student and teacher workshops, field trips, fossil cleaning demonstrations, and cooking lessons where participants made sweets resembling geological structures.

AAPG volunteers held talks at the General Lázaro Cárdenas School in **Mexico and Colombia**. The talks, World of Drilling and World of Rocks, were designed to help children fall in love with Earth sciences while learning about careers and natural resources.

Students and Young Professionals (YPs) in AAPG's Latin America and Caribbean Region implemented a region-wide strategy modeled after the Earth Science Week. Local communities in the region are participating in Earth-themed activities to promote awareness of the science. AAPG Student and Young Professional Chapters started organizing activities in Colombia, Brazil and Trinidad and Tobago.

AAPG Student and YP Chapters have partnered with the **Geological Society of Trinidad and Tobago (GSTT), the University of the West Indies, the SPE Trinidad and Tobago Section and private companies** to organize dozens of contests, field trips and lectures. Going with 2018's theme of "Earth as Inspiration," AAPG and GSTT organized an art, photography and videography competition themed around geoscience and Trinidad and Tobago culture. Educational Outreach in the country continues through a series of STEM (Science, Technology, Engineering and Math) lectures organized at high schools throughout the country. AAPG YP Chapter members provide talks for the lectures organized by Shell and Sacoda Serv.

Colombia, with the help of AAPG, expanded its Earth Science Week efforts in 2018, holding events in Barranquilla, Cúcuta, Manizales, Medellín and Valledupar, cities whose universities have geology programs. University students and professors organized local events and received support from the Colombian Geological Survey, the Colombian Geological Society and the Colombian Association of Geologists and Geophysicists.

Colombia also saw the second annual Earth Science Week Celebration at the University of Pamplona in the Cúcuta. The event brought 120 high school students from three schools to the campus. Participants attended a workshop on "Geology in your Cell Phone," which focused on the importance of geology and minerals in daily life. Students visited stands with hands-on activities focused on geological time, Earth resources, climate, energy and resources and risk management. Event volunteers hoped to encourage participants study geology. Students from the university also received training on how to communicate scientific knowledge to the public.

The first-ever AAPG **Geosciences Week** impacted 3,000 students and 100 teachers in **Mexico City**. The AAPG Mexico YP Chapter organized the event along with student volunteers from the National Polytechnic Institute (IPN) and the National Autonomous University of Mexico (UNAM) and professionals with expertise in micropaleontology, drilling, renewable energies and science communication. The Mexican Geological Survey donated a rock collection for the event, UNAM loaned a mobile geology laboratory, and IPN provided educational materials. Students concluded the week with a visit to the UNAM Geology Museum, home to one of Mexico's most important mineral collections, and they enjoyed a performance from the National Institute of Fine Arts orchestra. Geoscience Week Mexico also led to a signed agreement with PIXELDEMIA, a mobile application development company, who agreed to create a free app to provide geoscience education to children. Organizers plan to include the app in activities next year.

Activities are ongoing in Brazil, Peru, and Trinidad and Tobago. Earth Science Week topics are being discussed in universities in Argentina and Costa Rica. The newly announced 2019 theme of "Geoscience Is For Everyone," along with the release of the Online Toolkit, makes geoscience learning more accessible to a wider audience.

Earth Science Week Toolkits

AGI assembled some **14,000 Earth Science Week Toolkits,** the vast majority of which were distributed to teachers and geoscientists before the end of 2018. Twelve AGI member societies requested complimentary Earth Science Week Toolkits for distribution, and 30 state geological surveys requested complimentary kits for distribution.

As in past years, thousands of kits also were distributed through program partners including USGS, NASA, the National Park Service, and AAPG Student Chapters. Hundreds of kits were shipped free to geoscience department chairs at colleges and universities nationwide. Toolkits were shipped to program participants and around the world.

The 2018 toolkit featured AGI's traditional Earth Science Week poster, education and outreach flyer, and school-year calendar showcasing geoscience classroom investigations and important dates of Earth science events.

Continuing 2017's trend, the 2018–19 school-year calendar's classroom investigations featured notations explaining to educators exactly how each activity aligns with expectations outlined in the *Next Generation Science*

Earth Science Week 2018 Toolkit envelope. Artwork by AGI/K. Cantner.

Standards. Additionally, program partners' contributions, many of which also included activities aligned with the standards, made the Earth Science Week 2018 Toolkit one of the richest in recent years.

The **Online Toolkit** was also launched this year, including downloadable resources from some portions of the kit. Users can browse resources by theme/year, type of resource, and *Next Generation Science Standards* (NGSS) topics. The Online Toolkit houses resources dating back to 2014 and is a feasible option to organizations that want to lower their paper use, as well as users who are limited by shipping and printing costs. The online toolkit has gotten nearly 4,000 page views in the first three months (October to December 2018).

Web Resources

According to Google Analytics, the **Earth Science Week website** was accessed by users in **212 nations, territories, and regions** worldwide in 2018. The program website (**www.earthsciweek.org**) delivers essential resources for educators throughout the year. As in past years, the Earth Science Week website was updated regularly to reflect the new theme, contests, proclamations, events, initiatives, and classroom activities for 2018. The entire site received more than 550,000 page views. Within the site, Classroom Activities pages received over 342,000 views. Contests pages received nearly 40,000 views. Plan an Event pages received over 7,150 views.

Those hosting events during Earth Science Week 2018 were invited to let people know about it at **Events in Your Area.** This web page provided information on events taking place through program partners in each state, such as exhibits, tours, lectures, and open houses. The new **Earth Science Week Event Registry** enabled participants to promote their events more effectively than ever. All registered events were listed on Earth Science Week's **Events In Your Area** site.

In addition, participating groups could be listed in **Earth Science Organizations,** an online map that offers clickable links to Earth Science Week events taking place at parks, museums, science and technology centers, university geology departments, local geological societies, and other nearby locations.

Promoted through various online channels, AGI's **Earth Science Week promotional video** trumpeted the importance of the geosciences and the celebration's role in promoting public awareness. This brief, exciting, eye-popping video answers key questions: Why is Earth science a big deal? How does Earth Science Week help promote learning and teaching about the subject? And what can students, educators, community partners, and others do to get involved?

While exploring the Earth Science Week theme of "Earth as Inspiration," science teachers and students were invited to consider the ways that people showcase the Earth and its systems into other artful mediums. The **Our Shared Geoheritage** and **Visualizing Earth Systems** pages on the Earth Science Week website continue to be popular resources on the Earth Science Week website. Our Geological Heritage features educational material on this heritage and links users to recommended resources, including downloadable reports, articles, blogs, geoheritage locations, and learning activities. The page also features geoheritage-related classroom activities and links to information on geoheritage in

Earth Science Week 2018 Photo Contest entry by finalist Mary Francis Garcia.

every state. Visualizing Earth Systems links users to dozens of recommended visualizations dealing with energy, climate, minerals, water, hazards, and other topics. In addition, the page links users to overviews of these topics provided by AGI's Critical Issues Program.

Once again AGI offered four quarterly **Earth Science Week Webcasts** in 2018, expanding the program's use of online formats and media for public outreach. The free webcasts provided lively overviews of Focus Days (spring), Contests (summer), the Toolkit (autumn), and the Roy Award (winter). Each roughly five-minute tutorial includes a wealth of online links, which viewers can click during the narrated presentation to review available resources.

Program participants were encouraged to visit the continually updated Earth Science Week **Classroom Activities** page for more than 130 free learning activities, most of them contributed by leading geoscience agencies and groups. Activities are organized and searchable by various criteria, including specific Earth science topics. To find the perfect activity for a lesson, teachers can search by grade level and science education standard. Maybe most useful, they also can search among 17 categories of Earth science topics, from energy and environment to plate tectonics and weathering. This may be why Classroom Activities rank as one of the program's most popular online offerings, with 100 percent of survey respondents rating it as "useful" or "very useful."

AGI provided a set of free online videos and other electronic resources to help students, educators, and others explore the "big ideas" of Earth science during Earth Science Week 2018 and throughout the year. **Big Ideas Videos** bring to life the nine core geoscience concepts

Earth Science Week 2018 Visual Arts Contest entry by finalist Janiru Sumanasiri.

that everyone should know. The Earth Science Literacy Initiative, funded by the National Science Foundation, codified these principles. The videos are available on YouTube and TeacherTube. The Earth Science Week website also provides dozens of classroom activities linked to the "big ideas."

A page dedicated to **Geoscience Career, Scholarship, and Internship Resources** remains on the program website. Another page of links includes external connections to sites featuring resources on key topics such as chemistry, climate, drought, earthquakes, energy, floods, hurricanes, landslides, sinkholes, soil, tornadoes, tsunamis, volcanoes, and wilderness fires.

Finally, Earth Science Week makes ample use of online social networking to reach new audiences, especially young people. The program's presence on **Facebook,** the Internet's most popular networking site, includes an Earth Science Week Fan Page. In addition, web surfers are invited to receive geoscience news, resources, and opportunities by following Earth Science Week on **Twitter.** Tweets are sent frequently, whenever there is valuable news or information to share. The number of people learning about Earth Science Week through social media remained impressive in 2018, as more than 155,000 people received program information from AGI and program partners through Facebook, Twitter, and Pinterest.

Newsletter

The monthly **Earth Science Week Update** newsletter reached some 5,000 teacher, student, and geoscientist subscribers in the past year. The electronic newsletter kept planners and participants up-to-date on Earth Science Week planning at the national level, encouraged participation in local areas, and provided news on geoscience topics of interest to participants.

Besides highlighting worthwhile resources, these monthly e-mail updates reinforce the belief that

Earth Science Week 2018 Photo Contest finalist entry by Aaliyah Craven.

geoscience education is a priority throughout the year, not only for one week each October. It is little wonder that the e-newsletter remains a popular online offering, with 91 percent of survey respondents rating it as "useful" or "very useful."

Earth Science Week 2018 Visual Arts Contest entry by the winner, Namit Vernekar.

Contests

AGI held contests in connection with Earth Science Week for the 18th consecutive year. Contests were designed to encourage K–12 students, teachers, and the general public to become involved in the celebration by exploring artistic and academic applications of Earth science. Earth Science Week continued expanded eligibility for its photo contest to allow international members of **AGI Member Societies** and **AGI International Affiliates** to participate.

Four contests continued to provide ways for many people to participate in Earth Science Week. Photos, art, videos, and essays were submitted by hundreds of people. Each first-place winner received $300 and a copy of AGI's The Geoscience Handbook. Entries submitted by winners and finalists were posted online.

Namit Vernekar of Charlotte, North Carolina, won first place in the visual arts contest with a creative and colorful drawing of two sides (real and imaginary) of Earth. Finalists were Janiru Sumanasiri, Esther Gammill, Saachi Tamboli, and Abhiraj Das. Students in grades K–5 made two-dimensional artworks illustrating the theme "**Earth and Art.**" Matt Meisenheimer of Janesville, Wisconsin, won first place in the photo contest with an image of Kalalau Valley in Kauai, Hawaii. Finalists were Hannah Kawar, Aaliyah Craven, Mary Francis Garcia, and Kate Ragusa. Submissions illustrated the theme "**Inspired by Earth.**" Udbhav Akolkar of Scottsdale, Arizona, won first place in the essay contest with a paper on Earth in various forms of writing, from J. R. R. Tolken Edgar Allan Poe. Finalists were Caden Longwater, Vikram Kolli, Jaeho Lee, and Lluvia Perez. Students in

Earth Science Week 2018 Photo Contest entry by Hannah Kawar.

Below, part of the winning entry to AGI's 2018 Earth Science Week essay contest by Udbhav Akolkar. See all the winners and read the rest of Udbhav's essay at **https://www.earthsciweek.org/contests/2018**

Earth as Inspiration
by Udbhav Akolkar

"Fire belched from its riven summit. The skies burst into thunder seared with lightning…." what would "Lord of the Rings" be without the description of magnificent mountains, raging rivers, dark forests and green meadows. All through ages, the art of writing has been inspired by Earth and its beauty, power and destructive forces.

The tales of endeavors through great peril and difficulty often show mountains as a major obstacle to overcome. The mountain ranges seem to loom over the rest of the land, foreshadowing the doom. It also is an opportunity to demonstrate the bravado of their characters.

grades 6–9 wrote essays of up to 300 words addressing this year's theme, "**Finding 'Art' in Earth**." Noah Resnik won first place in the video contest with a public service announcement about ways science (and the Earth) can be artful. Finalists were (team) Carrie Hunter and Austin Hermann; (team) Zihao Jiang, Chen Hong Xue, Yu Hongyi, Lu Yichang, Ding Keyi, Liu Zhengyi, Wang Yihan, Zhang Xin Yao, Ding Zi Jun, Chen Yan Yan, Hu Rui, Yin Hao Ran, Cao Yi Jia, Wu Di, and Sun Dasheng; (team) Vinuth and Janiru Sumanasiri and Malika Gunasekara; and (team) Rossalyn Buck and Emily Straetz. Individuals and teams created brief, original videos that tell viewers how artistic expression stems from the natural world, through "**Earth Expressions**" in their part of the world.

In cooperation with Washington, D.C.'s Union Station, Mazza Gallerie, and the George Bush Intercontinental Airport, Earth Science Week staged an "Earth and Human Activity Here" Photo Exhibit featuring select images from in the 2017 photo contest. Union Station's historic facility is visited by some 100,000 train and local metro passengers passing through daily, while Mazza Gallerie attracts thousands of shoppers every year. The George Bush Intercontinental Airport is the 15th busiest in the nation, seeing over 110,000 travelers daily. The partnership launched what is expected to be an ongoing program offering, exhibits of photos in high-traffic public spaces nationwide.

Earth Science Teacher Award

For the 11th consecutive year, AGI and the AGI Foundation offered the **Edward C. Roy, Jr. Award for Excellence in K–8 Earth Science Teaching.** The 2018 award went to **Kenneth L. Huff,** a sixth-grade teacher at Mill Middle School in Williamsville, New York. Huff earned his bachelor's and master's degrees from the State University of New York College at Buffalo. He is a member of several professional organizations, including the Science Teachers Association of New York State, and is the recipient of numerous awards for his work in education.

Huff received a $2,500 prize and an additional grant of $1,000 to enable him to attend the National Science Teachers Association 2018 National Conference to accept the award during a reception hosted by the National Earth Science Teachers Association. Finalists for the award were Anica Brown of Pound Middle School in Lincoln, Nebraska, and Christopher Spiegl of Montgomery Bell Academy in Nashville, Tennessee.

The award recognizes one classroom teacher from kindergarten to eighth grade for leadership and innovation in Earth science education. This award is named in honor of Dr. Edward C. Roy, Jr., a past president of AGI and

2018 Edward C. Roy, Jr. Award winner Kenneth L. Huff and Juliet Crowell at the NESTA reception during NSTA's 2018 National Conference.
Image credit: Tom Ervin

strong supporter of Earth science education. In addition to U.S. teachers, instructors throughout the United Kingdom were invited to compete for the prize. U.K. teachers were provided with detailed guidance on entering the competition by AGI and **The Geological Society of London**, a member society and Earth Science Week partner.

Focus Days

Earth Science Week 2018 kicked off on Sunday, October 14, with the 10th annual **International EarthCache Day**. "EarthCaching" is a variation of a recreational activity known as geocaching, in which a geocache organizer posts latitude and longitude coordinates on the Internet to advertise a cache that geocachers locate using GPS devices. The activity has attracted over a

million participants worldwide. When people visit an EarthCache, they learn something special about Earth science, the geology of the location, or how the Earth's resources and environment are managed there. EarthCaching has been developed by the Geological Society of America, a major program partner.

On Monday, October 15, educators and young people were encouraged to explore "big ideas" as part of **Earth Science Literacy Day.** The AGI "Big Ideas of Earth Science" videos provided on YouTube and TeacherTube outline the core concepts of geoscience, as codified by the Earth Science Literacy Initiative with support from the National Science Foundation. To help teachers and students use the videos, the Earth Science Week website offers dozens of related classroom activities.

One of the highlights of recent years' Earth Science Week celebrations has been **"No Child Left Inside" Day,** an event that in its inaugural year engaged some 500 students in outdoor learning activities and received coverage by news media from NBC to NPR. In 2018, students and educators nationwide were invited to take part on the Tuesday of Earth Science Week, October 16. AGI's online NCLI Day Guide provided everything needed to plan a local NCLI Day event. The free guide provided 17 outdoor activities, as well as detailed recommendations for creating partnerships, planning logistics, reaching out to the local media, and following up in the classroom.

In addition, on the Tuesday of Earth Science Week 2018 participants were invited to take part in **Earth Observation Day.** Previously celebrated at other times of the year, this October 16 event aimed to engage students and teachers in remote sensing as an exciting and powerful educational tool. The event was a STEM educational outreach event of AmericaView and its partners. AmericaView is a nationwide partnership of remote sensing scientists who support the use of Landsat and other public domain remotely sensed satellite data through applied remote sensing research, K–12 and higher STEM education, workforce development, and technology transfer. Participants made use of lessons and activities by AmericaView and other organizations, as well as additional Earth Observation Day resources, online.

Earth Science featured the return of a popular event, **National Fossil Day**. In partnership with the National Park Service (NPS), AGI helped conduct the ninth annual event, officially on Wednesday, October 17, 2018 including activities and resources designed to celebrate the scientific and educational value of fossils, paleontology, and the importance of preserving fossils for future generations. NPS offered a website full of educational resources and information designed specifically for students and teachers. On the site's NPS Fossil Park Highlights page, visitors could find lesson plans developed to reflect state standards, fossil trading cards, videos about pygmy mammoths, special brochures, a virtual museum exhibit on dinosaurs, and more. NPS also held a National Fossil Day Art Contest. Finally, AGI

The "Green Screen" activity at AGI's booth during the National Fossil Day event.
Image credit: NPS/Anthony DeYoung

collaborated with NPS partners and other geoscience organizations to conduct a National Fossil Day event in Washington, D.C., held the week before on Wednesday, October 3, 2018. Fossil enthusiasts in Washington, DC celebrated National Fossil Day on the National Mall from, where AGI educated and entertained visitors with a green screen-equipped "Paleontology Play Space" photo booth, funded by the Paleontological Society, on the pathway between the Smithsonian Castle and the National Museum of Natural History. Visitors, who had their pictures taken virtually in the midst of amazing fossil finds, received photo souvenirs of the day.

Program participants were invited to join the Earth Science Week team in encouraging everyone — including women, minorities, and people with a range of abilities — to explore geoscience careers on **Geoscience for Everyone Day,** Thursday, October 18. Educators welcomed geoscientists into the classroom to speak. Geoscientists visited schools and volunteered at science centers. Others organized scout events, led field trips, and held special "Take Your Child to Work Day" events. The aim was to open a young person's eyes to the world of Earth science. Doing so, participants supported the efforts of AGI member societies such as the Association for Women Geoscientists and the National Association of Black Geoscientists in raising awareness of the remarkable opportunities available to all young people in the Earth sciences. The program website directed participants to "Visiting Geoscientists: An Outreach Guide for Geoscience Professionals," a handbook co-produced by AGI and the American Association of Petroleum Geologists' Youth Education Activities Committee.

The seventh annual **Geologic Map Day** held on Friday, October 19, 2018, promoted awareness of the study, uses, and importance of geologic mapping for education, science, business, and a variety of public policy concerns. The final event for the school week of Earth Science Week 2018 was hosted by the U.S. Geological Survey and the Association of American State Geologists in partnership with AGI, along with additional partners including the National Park Service, the Geological Society of America, and NASA. Students, teachers, and the wider public tapped into the various educational activities, print materials, online resources, and public outreach opportunities for active participation. The Earth Science Week 2018 Toolkit contained a Geologic Map Day poster that provided geologic maps, plus step-by-step instructions for a related classroom activity on artistic inspiration. Additional resources for learning about geologic maps were featured on the Geologic Map Day web page of the Earth Science Week

site. Activities nationwide, many led by state geologic surveys, spurred learning in schools.

Earth Science Week 2018 reached its climax with **International Archaeology Day** on Saturday, October 20. The event was a celebration of archaeology and the thrill of discovery. Every October, archaeological programs and activities for people of all ages and interests are presented by the Archaeological Institute of America and archaeological organizations across the United States, Canada, and elsewhere. Programs included activities such as a family-friendly archaeology fair, a guided tour of a local archaeological site, a simulated dig, and a lecture or a classroom visit from an archaeologist. In every case, interactive, hands-on International Archaeology Day programs provided the chance for participants to indulge their inner "Indiana Jones."

Special Events

City-specific celebrations served as major centers of public awareness activities during Earth Science Week 2018. **Houston and Denver** extended and deepened the reach of the nationwide event.

AGI's **Citywide Celebrations website** provided educators, students, and families with links to additional educational resources as well as other offerings in participating cities. Program participants nationwide were encouraged to collaborate with local partners to launch their own Citywide Celebration.

AGI staff exhibited at the **Energy Day Houston** festival as part of its new collaboration with Consumer Energy Alliance. Earth Science Week shared geoscience-oriented STEM materials with the many of people who visited its booth at this free outdoor event, which for years has showcased education technologies and innovations. Organizers estimated roughly 25,000 people attended the event.

Earth Science Week 2018 opened and closed with local events at Alexandria's Torpedo Factory Arts Center. The kickoff event for 2018's "Earth As Inspiration" theme welcomed hundreds of educators, students, and the general public to explore connections between the geosciences and the arts during **"The Late Shift: STEAM-Powered December."** The free event included hands-on activities, demonstrations linking art and geoscience, musical performances, and opportunities to craft "take home" artworks based in Earth science. Additionally, educational materials such as Earth Science Week Toolkits were distributed to teachers, students,

AGI's John Northington and Sequoyah McGee at 2018 Energy Day Festival in Houston, Texas.
Image credit: CEA/Kathleen Koehler van Keppel

homeschoolers, and others. Roy Award winner M.J. Tykoski was formally recognized for excellence in Earth science teaching. A highlight of the evening was a series of addresses on the importance of geoscience by **AGI Earth Science Education Ambassador and former U.S. Secretary of the Interior Sally Jewell, Alexandria Mayor Allison Silberberg, and AGI Executive Director Allyson Anderson Book.**

Earth Science Week 2018 again amplified their engagement with a pre-Earth Science Week appearance at the Torpedo Factory's annual **Art Safari**. AGI staff, along

Participant in 2018 STEAM workshop led by AGI's education staff.
Image credit: AGI/Celia Payne

with Torpedo Factory artists, facilitated hands-on art activities for the public. The AGI/Earth Science Week booths created colorful tracings of dinosaurs on rainbow scratch paper while distributing dinosaur- and fossil-specific materials, such as posters and activities. Earth Science Week also facilitated a scavenger hunt that took visitors through the Torpedo Factory, highlighting the geoscience behind art exhibits, natural phenomena, and the building's structure itself. Winners of the scavenger hunt were given a prize fossil, mineral, or rock kit with links to educational activities on the website. In total, an estimated 500 visitors participated in the Art Safari event.

AGI Promotions

Earth Science Week promoted awareness of numerous AGI programs and resources of interest to Earth science educators, students, and enthusiasts, including AGI's Center for Geoscience & Society, the AGI Geoscience Workforce program, the William L. Fisher Congressional Geoscience Fellowship, AGI's NGSS Education Webinars, the Pulse of Earth Science website, the Visiting Geoscientists guide, AGI's Critical Issues Program, Geoscience in Your State Factsheets, the Earth Science Organizations website, the Faces of Earth DVD, EARTH magazine, AGI Information Services (such as GeoRef), District Visit Days, the Why Earth Science video, and AGI's and the National Park Service's jointly published America's Geologic Heritage: An Invitation to Leadership.

Congressional Recognition

In a display of bipartisanship, members of the U.S. House of Representatives from across the country came together to introduce a resolution expressing support for designation of the week of October 14–20, 2018, as Earth Science Week.

Representative Jared Polis (D-Colorado) submitted House Resolution 556, on behalf of himself and Representatives Barbara Comstock (R-Virginia), Dan Lipinski (D-Illinois), and Ryan Costello (R-Pennsylvania) on October 5, 2018. Recognizing Earth Science Week 2018 as the 21st annual celebration of the signature event organized by the AGI, the resolution noted that "the study of Earth sciences leads to an improved understanding of the Earth's natural systems and the interplay between human society and these systems."

State Proclamations

Seven states have demonstrated outstanding science-literacy leadership by issuing **"perpetual proclamations"** of Earth Science Week, ensuring recognition every year: Alaska, Delaware, Illinois, Nevada, North Dakota, Oklahoma, and South Dakota.

Governors also issued single-year proclamations in additional states — Alabama, Arkansas, Missouri, New Jersey, Pennsylvania, and Tennessee — bringing the total number of states with proclamations of recognizing Earth Science Week 2018 to 13.

Publicity and Media Coverage

AGI enlisted the support of a wide range of media in promoting awareness of Earth Science Week, resulting in unprecedented reach for promotional activities in 2018 and helping to lay a foundation for more coverage in years to come.

Earth Science Week 2018 news, events, programs, and resources were covered by **national news organizations** such as the American Anthropological Association, American Geophysical union, Archaeological Institute of America, AAPG Explorer, EventBrite, Facebook, Geocaching, Instagram, National Park Service, National Science Teachers Association Book Beats, Pinterest, Public Library of Science (PLOS), Reddit, SERC Media, Smithsonian National Museum of Natural History, Society of Exploration Geophysicists, TeachersFirst, and Zulily.

Additionally, the event was covered by **international news organizations** including Bharath Gyan, Bivash Vlogs, Breaking Belize News, CHON-FM 98.1; Dart Connections, Earth Science Week Japan in Shizuoka, Education HQ-Australia, Geological Society of London and Geological Society of London Blog, International Heritage News Network, International Paleontological Association, Jividha, Laboratory News, On The Wight, PuneMirror, The Courier-Glasgow, The Guardian, The Tribune-India, and ZAWYA.

Throughout the United States, coverage of Earth Science Week programs and activities was provided by **local news organizations** such as 8th Streel Ale Haus of Sheboygan, Wisconsin; AdVantage News of Illinois; Alignable; Allevents.in; Alton Daily News of Illinois; Bakersfield and Bakersfield Now of Bakersfield, California; Before It's News; Best Things Florida; Best Things Georgia; Bham Now of Birmingham Alabama; Birmingham 365 of Birmingham, Alabama; Bluff Country Newspaper Group of Spring Valley, Minnesota; BoZone of Bozeman, Montana; Clay Today of Clay County, Florida; Click 2 Houston of Houston, Texas; Coast Weekend of Oregon; Columbus Underground of Columbus Ohio; Commonwealth Heritage Group of Dexter, Michigan; County 10 and County 17 of Wyoming; Culture Map Houston of Texas; Delaware Museum of Natural History; Democratic Underground of Kensington, Maryland; Dickinson; Digital Journal; Discovery Park of America in Obion County, Tennessee; DNA Tube; Do512 of Austin, Texas; Downtown Indy of Indianapolis, Indiana; Earth Science Club of Northern Illinois; El Paso Herald-Post of Texas; Environmental Education in Tennessee; Evensi; Fairbanks Daily News-Miner of Fairbanks, Alaska; Fallon County Times of Baker, Montana; Falls of the Ohio State Park of Clarksville, Indiana; Filmz Video; Florida Public Archaeology Network of Pensacola, Florida; GeoTripper; GetLink; Global Mining Review; Hays Post of Hays, Kansas; Helena High School of Helena, Montana; Hernando Sun of Hernando County, Florida; Hometown Focus of Virginia, Minnesota; Hoo Knows; Hudson Star-Observer of River Falls, Wisconsin; Indy Arts Guide of Indianapolis, Indiana; Inside CSUSB of San Bernadino, California; Ithaca Times of Ithaca, New York; Javelinas Collegiate Link of Texas; Johnson City Press of Johnson City, Tennessee; Jumpic.com; Kenosha News of Kenosha, Wisconsin; KC STEM Alliance of Kansas City, Missouri; Las Cruces Sun News of Las Cruces, New Mexico; LearningMagazine.com; Los Alamos Daily Post of Los Alamos, New Mexico; Macaroni Kid; Martinsville Bulletin of Martinsville, Virginia; Mesa Public Library of Mesa, Arizona; MiningNews.net; Missouri Department of Natural Resources; Mitchell South Dakota; Montana Untamed; MOREnet; Mother Nature Network; Moultrie News of South Carolina; Museum of the Rockies of Bozeman, Montana; MV Times of Martha's Vineyard, Massachusetts; Napa Valley Register of California; Nevada Bureau of Mines and Geology Blog; News Locker; Newstral; News Tribune of Jefferson City, Missouri; NWTN Today of Tennessee; Ozobot; Payson Roundup of Payson, Arizona; Penn State; PennState Brandywine; Phoenix New Times of Phoenix, Arizona; Platte County Record-Times of Wyoming; Providence Daily Dose of Providence, Rhode Island; RiverBender.com; Rosie on the House of Phoenix, Arizona; RRSTAR of Rockford, Illinois; Rutherford Source of Rutherford County, Tennessee; San Jacinto Times of Shepherd, Texas; SanJac Watercooler Employee Newsletter; Santa Barbara Independent of Santa Barbara, California; Serendeputy; sNEWSi; Star Tribune Casper of Casper, Wyoming; State Gazette of Dyersburg, Tennessee; St. Augustine Lighthouse & Maritime Museum of St. Augustine, Florida; Stillwater News of Stillwater, Oklahoma; Super News World; SYS-Con of New Jersey; Tallahassee Democrat of Tallahassee, Florida; Texas Geosciences of The University of Texas at Austin, Jackson School of Geosciences; Texas Parks & Wildlife; The Bengal of Pocatello, Idaho; The Brunswick of Brunswick, Georgia; The Bulletin of Norwich, Connecticut; The City of Fargo, North Dakota; The Courier Waterloo-Cedar Falls, of Waterloo, Iowa; The Crimson-White of Alabama; The Daily Camera of Boulder, Colorado; The Denver Post of Colorado; The Gainsville Sun of Gainsville, Florida; The Gazette of Colorado Springs, Colorado; The Greeneville Sun of Greeneville, Tennessee; The Herald-Dispatch of Huntington, West Virginia; The Indiana Gazette; The Jamestown Sun of Jamestown, North Dakota; The Kemmerer Gazette of Lincoln County, Wyoming; The Knoxville Focus of Tennessee; The Lowell Sun of Lowell,

Massachusetts; The Mining Journal of Marquette, Michigan; The Missourian; The Montana Dinosaur Trail; The Newtown Bee of Newtown Connecticut; The Northwest Arkansas Democrat-Gazette; The Ocala Star Banner of Ocala, Florida; The Ohio Earth Scientist; The Oneida Daily Dispatch of Oneida, New York; The Paris News of Paris, Texas; The Pine Log of Nacogdoches, Texas; The Pitch of Kansas City, Missouri; The Press-Enterprise of Riverside, California; The Sand Mountain Reporter of Albertville, Alabama; The Stann Creek Regional Archaeology Project; The State Museum of Pennsylvania; The Sun Times News of Chelsea, Michigan; The University of Memphis in Tennessee; The University of Tennessee; The University of Tennessee, Knoxville; The University of Oregon Events; The University of Wisconsin, Milwaukee; The Utah Geological Survey; The Villager; Time.ly; Time Scavengers; Topix; TV6 Upper Michigan Source; University of Kentucky; University of Nevada, Reno; University of Wyoming UW News; UpMatters; VidByte; WDTN.com of Dayton, Ohio; Weekly Villager of Garrettsville, Ohio; West Kentucky Star; Western Slope Now; and Winona Daily News of Winona, Minnesota.

Earth Science Week also was covered by **television and radio stations** across the country, including KPVI News 6 and Local News 8 in Idaho; ABC 13 in Texas; ABC 15 in Arizona; CBS Denver; FOX 10 News of Alabama; FOX47 News of Michigan; K2 Radio of Wyoming; Kansas Public Radio; KKCO-TV of Colorado; KALW of California; KNAU Arizona Public Radio; KNOE NEws

Participant making paper in an AGI 2018 STEAM workshop.
Image credit: AGI/ Celia Payne

of Louisiana; KRQE of New Mexico; WGNS News Radio of Tennessee; WTVG TV of Ohio; WVLT8-TV of Tennessee; and YxtMp3.live.

AGI distributed **press releases** to hundreds of newspapers, magazines, and other print media outlets. The articles highlighted Earth Science Week activities and the program theme. Press releases about Earth Science Week activities were also released by Mountain Xpress, PR.com, and Wyoming State Geological Survey.

Copies of the **Earth Science Week 2018 poster,** featuring a geoscience learning activity in addition to promotional content, were distributed as inserts in publications carrying articles about the event, such as NESTA's The Earth Scientist, GSA Today, AAPG Explorer, and AGI's EARTH magazine. In addition, the poster image appeared in the Society of Exploration Geoscientists' The Leading Edge, which reaches some 100,000 people in print and online.

External Evaluation of Earth Science Week 2018: Key Findings

Following the event, AGI secured an independent contractor, PS International, to complete a formal external evaluation of Earth Science Week 2018, as it has in past years. Participants were invited to participate in a survey in the closing months of 2018, with a valid response rate of 6.5 percent.

Results were overwhelmingly positive. Participation remained strong. Comparing participation last year and plans for next year, 99 percent of survey respondents said they anticipate either increasing or maintaining level participation.

A large majority (92 percent) said Earth Science Week offers opportunities for teaching and promoting Earth science that they would not have otherwise. Similarly, 90 percent said program resources and activities are very or somewhat important to educating students about geoscience.

Eighty-seven percent of respondents rated the program's overall usefulness as "excellent" or "good." When respondents rated nearly 20 key items from the Earth Science Week Toolkit and Website — such as posters, disks, and online activities — all were rated "very useful" or "useful" by strong majorities of participants.

Participants said they were active during Earth Science Week. Many reported specific activities that were highly active. For example, 89 percent reported activities categorized as "most active" (e.g., field trips and outside lessons), "active" (e.g., external speakers and open house discussions), or "somewhat active" (e.g., lesson plans and kit distribution).

Asked how Earth Science Week might be improved, respondents advocated additional program materials and activities, as well as increasing communication and promotion. AGI uses these evaluation findings to fuel continual improvement of the program.

Earth Science Week Sponsors

United States Geological Survey	Howard Hughes Medical Institute	Minerals Education Coalition
National Aeronautics and Space Administration	Association of American State Geologists	Schlumberger
National Park Service	Geological Society of America	Water Footprint Calculator (Grace Communications Foundation)
American Association of Petroleum Geologists Foundation	Archaeological Institute of America	Consumer Energy Education Foundation
American Geophysical Union	AmericaView	Keystone Policy Center
Society for Mining, Metallurgy, and Exploration	Equinor	TGS

Earth Science Week Program Partners

American Association of Petroleum Geologists	Equinor	Student Energy
American Association of Petroleum Geologists Foundation	Geological Society of America	Switch Energy Project
	Google	TERC
American Geophysical Union	Howard Hughes Medical Institute	TGS
American Geosciences Institute	Incorporated Research Institutions for Seismology	Torpedo Factory Art Center
American Meteorological Society	Keystone Policy Center	UNAVCO
AmericaView	Minerals Education Coalition	U.S. Bureau of Land Management
Archaeological Institute of America	National Earth Science Teachers Association	U.S. Geological Survey
Association of American State Geologists	Science Friday	U.S. National Aeronautics and Space Administration
California Geological Survey	Schlumberger	U.S. National Oceanic and Atmospheric Administration
CLEAN Network	Society of Exploration Geophysicists	U.S. National Park Service
Consumer Energy Education Foundation	Soil Science Society of America	Water Footprint Calculator (Grace Communications Foundation)
Critical Zones Observatories	Society for Mining, Metallurgy, and Exploration	
EarthScope		

Earth Science Week 2018 Events and Activities by State and Territory

While it is impossible to track all Earth Science Week activities in the United States, major activities across the country included:

Alabama

- The Alabama Museum of Natural History hosted a celebration of all things fossil during National Fossil Day at the University of Alabama.
- The governor of Alabama issued a proclamation for Earth Science Week 2018.
- The Geological Survey of Alabama and the International Association of Hydrogeologists distributed complimentary Earth Science Week Toolkits to science educators for use in the classroom.
- Earth Science Week was promoted among teachers and students by the Environmental Education Association of Alabama.
- The University of South Alabama Archaeology Museum welcomed visitors of all ages to participate in International Archaeology Day on October 20. The event offered a variety of indoor and outdoor hands-on activities for visitors to try.
- The University of Alabama's Smith Hall Museum of Natural History hosted a National Fossil Day event celebrating fossil appreciation and stewardship. The public was invited to meet a paleontologist and view fossils.
- The North Alabama Society, along with other partners, including the Alabama Archaeological Society, sponsored an Archaeology Fair on Saturday, October 20th from 1:00–4:00 PM at Lowe Mill Dock and Lawn.

Alaska

- The governor of Alaska issued a perpetual proclamation of Earth Science Week.
- Guests were invited to join the Bureau of Land Management, Alaska and its partners to celebrate International Archaeology Day at the Campbell Creek Science Center.
- At the University of Alaska Museum of the North's "Ask an Archaeologist," visitors met museum archaeologists in the lobby from noon to 4 p.m. Oct. 1–5.

Arizona

- The Museum of Northern Arizona held a "Fossil Day" with exciting kids' programs, hands-on activities, and creative crafts.
- Grand Canyon National Park hosted multiple events for its annual Earth Science Week and National Fossil Day celebration. Activities included ranger-led hikes exploring geologic time and paleontology lectures.
- The Museum of Northern Arizona hosted a "Digging Dinosaurs" National Fossil Day celebration. Visitors were invited to touch fossils, handle tools and hear how paleontologists find, excavate, remove and prepare dinosaur bones.
- There was a celebration for National Fossil Day at Petrified Forest National Park in a unique Triassic style. Activities included watching fossil preparation, touring museum collections, creating plaster fossil casts, and talking with paleontologists.
- Udbhav Akolkar of Scottsdale, Arizona, won first place in the Earth Science Week essay contest with a paper on "Earth as Inspiration" found in various forms of writing.
- From 10:00 a.m. to 2:00 p.m. at the Visitor Center, the general public was invited to learn about the intricate pottery designs of the Salado from park archeologists during International Archaeology Day at Tonto National Monument.
- Geoscientists, with GeoWorld Travel, led a 10-day guided geology tour from Bozeman, Montana to Phoenix, Arizona with stops at four national parks, three national monuments and two world heritage sites.
- The Pueblo Grande Museum in Phoenix hosted an International Archeology Day event in collaboration with the Central Arizona Society of the Archeological Institute of America that featured archeological demonstrations, children's activities, tours, and more.
- Earth Science Week 2018 Toolkits were distributed to students, teachers, and others by Carla McAuliffe of the National Earth Science Teachers Association.

Arkansas

- The Arkansas Geological Survey promoted Earth Science Week among teachers and citizens with the distribution of related instructional materials.
- The governor of Arkansas issued a proclamation for Earth Science Week 2018.
- The Fall 2018 meeting of the South Central Historical Archaeology Conference was held in Arkadelphia October 26th – 28th, hosted by Henderson State University's Geography and Anthropology programs.
- Hands-on Archaeology Festival, a fun and educational hands-on archaeology event, was sponsored by University of Arkansas at Little Rock Anthropology

in collaboration with the Toltec Mounds Research Station of the Arkansas Archeological Survey and the Toltec Mounds Archeological State Park. While at the event visitors were invited to throw spears, make pottery, learn about ancient foodways, stone tool making demonstrations, tour a prehistoric mound group and more.

California

- The California Geological Survey distributed Earth Science Week Toolkits to educators.
- Finalists in the Earth Science Week photography contest included Mary Francis Garcia of San Carlos, Kate Ragusa of Redwood City, and Hannah Kawar of San Mateo.
- The Dr. John D. Cooper Archaeological and Paleontological Center celebrated International Archaeology Day and National Fossil Day with a free educational family event.
- The National Park Service hosted a National Fossil Day celebration at the Visitor Center at Joshua Tree National Park.
- The Raymond M. Alf Museum of Paleontology, The Western Science Center, and Cosplay for Science celebrated National Fossil Day at Los Angeles Comic Con. Paleontologists showed amazing fossils from their museums and talked about how these fossils helped to inspire some of our favorite monsters in movies, games, and comics.
- The Western Science Center and the Raymond M. Alf Museum of Paleontology celebrated both National Fossil Day and National Bison Day during "The Science of Fossils" with an up-close look at the bison that roamed our valley during the Ice Ages.
- San Diego Natural History Museum (The Nat) held a celebration of all things prehistoric for National Fossil Day with some show-and-tell and mini-tours.

- The Natural History Museum of Los Angeles County hosted a two-day festival celebrating dinosaurs for their 3rd annual Dino Fest during National Fossil Day.
- Nearly 300 people attended the International Archaeology Day and California Archaeology Month Expo held in Riverside. The archaeology fair included more than 15 exhibitors from academic, government, and private organizations. Attendees threw atlatl darts, identified artifacts, discovered ancient Egypt, experienced archaeological sites in virtual reality, and much more.

Colorado

- Aiming to get students interested in STEM careers, Consumer Energy Alliance and Consumer Energy Foundation partnered with dozens of organizations to host a free Energy Day Festival, featuring exciting energy related demonstrations and exhibits, at East High School.
- Royal Gorge Regional Museum and History Center hosted a Fossil Excavation Demonstration with Paleontologist Andrew Smith to celebrate National Fossil Day.
- The Minerals Education Coalition promoted Earth Science Week on their website.
- Colorado University Museum of Natural History invited the public to join them in celebrating National Fossil Day.
- In honor of Earth Science Week, two geology hikes were led by USGS geologist Pete Modreski—one up to the top of "Castle Rock" on South Table Mountain and another along the Green Mountain Trail.
- In conjunction with Girl Scout Day, Dinosaur Ridge Visitor Center hosted a National Fossil Day "Friends of Dinosaur Ridge" celebration.
- The Colorado Canyons Association hosted hikes to fossil sites

in the three Colorado National Conservation Areas for National Fossil Day.
- The Museums of Western Colorado held events featuring free educational activities for families across Colorado.
- Florissant Fossil Beds National Monument sponsored a day of fossil discovery including story time, fossil crafts, and a scavenger hunt.
- City of Colorado Springs and Western Interior Paleontological Society (WIPS) hosted a free family fun day with partners hosting educational tables, fossil identification, lots of hands-on demos, fossils and natural history giveaways, and free public lectures.
- Earth Science Week 2018 Toolkits were distributed to students, teachers, and others by representatives of the Colorado Geological Survey, the American Institute of Professional Geologists, the Geological Society of America, the National Earth Science Teachers Association, the Society for Mining, Metallurgy & Exploration, and the Society of Economic Geologists.

Connecticut

- EverWonder Children's Museum invited the public to come by on National Fossil Day to create their own fossils and explore other fossil activities.
- The Connecticut Geological and Natural History Survey promoted Earth Science Week and distributed kits among educators in the state.

Delaware

- The governor of Delaware issued a perpetual proclamation of Earth Science Week.
- The Delaware Geological Survey promoted Earth Science Week and Geologic Map Day by distributing Earth Science Week Toolkits and informational materials to educators.

• In addition, the Delaware Geological Survey recommended fossil hunters visit the Chesapeake and Delaware Canal to search for the state fossil, a belemnite, to celebrate National Fossil Day.

District of Columbia

• The National Park Service and AGI hosted exhibits at the National Fossil Day celebration on the National Mall.
• Numerous federal agencies, including the U.S. Geological Survey, NASA, and the National Park Service, supported Earth Science Week with the provision of teaching materials, distribution of kits, and events and activities.
• Earth Science Week 2018 Toolkits were distributed to students, teachers, and others by representatives of the American Geophysical Union.
• Earth Science Week was formally recognized in a resolution introduced in the U.S. House of Representatives.
• The National Aeronautics and Space Administration promoted the Earth Science Week 2018 photography contest on their website.
• The Geological Society of America encouraged geocachers around the world to participate in International EarthCache Day during Earth Science Week.
• The Bureau of Land Management (BLM) promoted the participation in Earth Science Week with emphasis on their participation in 2018 National Fossil Day (NFD) and International Archaeology Day (IAD).
• The National Education Association promoted Earth Science Week on their website.
• The National Museum of Natural History, the Society for American Archaeology, and Archaeology in the Community hosted a family-friendly event celebrating International Archaeology Day. There visitors could talk to archaeologists, hone their archaeology

skills, explore collections, and get free resources.

Florida

• The Florida Geological Survey promoted Earth Science Week by distributing educational materials.
• The 5th Annual Archaeologists for Autism was held in celebration of National Fossil Day to provide children and young adults with autism spectrum disorders and their families a chance to experience archaeology (and paleontology) in a fun, low stress environment. This event included fossil displays as well as a fossil "excavation pit."
• The Florida Geological Survey held an Earth Science Week Open House as well as a Wakulla Springs Field Trip promoting inspiration through exploration.
• To coincide with National Fossil Day, the Florida Museum of Natural History held a fun-for-all-ages celebration of its newest featured exhibit, Permian Monsters: Life Before the Dinosaurs, and explore the world before dinosaurs roamed the earth.

Georgia

• Visitors brought their own fossils to the William P. Wall Museum of Natural History at Georgia College for the museum's National Fossil Day ID and Presentation event.
• Students at Wilkes Primary School in Washington held a No Child Left Inside Day event sponsored by the Iris Garden Club, featured various outdoor learning activities guided by local experts in soil, forestry, music, art, drama, and nature.
• In celebration of National Fossil Day, preschoolers were invited to enjoy a story and special activity with a Fernbank Museum of Natural History educator featuring Bones, Bones, Dinosaur Bones by Byron Barton.

• Ocumulgee National Monument invited visitors to join them in celebration of National Fossil Day.

Hawaii

• GEOetc offered a trip to Hawaii's active volcanoes for STEM & Geography Teachers.

Idaho

• Idaho Museum of Natural History celebrated Earth Science Week and International Archaeology Day by sponsoring a day of hands-on, family-friendly archaeology activities.
• The Field Museum offered a unique tour developed specially to celebrate National Fossil Day.
• A celebration of National Fossil Day and Earth Science Week was held at the Idaho Museum of Mining and Geology in collaboration with the Orma J. Smith Museum of Natural History. Visitors could "explore Idaho during the Mesozoic" with activities about Idaho's geology, fossils, and dinosaurs.
• The Idaho Museum of Natural History hosted a family event for National Fossil Day. Visitors were invited to help the museum sort through material containing fossils.
• Burpee Museum's National Fossil Day 2018 programs included local fossil collectors and exhibitors, hands-on fossil activities, screening for microvertebrates, fossil ID's, Jane's REAL skull, and more.
• The Idaho Geological Survey distributed Earth Science Week educational materials.

Illinois

• The governor of Illinois issued a perpetual proclamation of Earth Science Week.
• Burpee Museum's National Fossil Day programs included local fossil collectors and various exhibitors' hands-on fossil activities.

- A special guided tour was held through the halls of the Field Museum where visitors learned about the fascinating story of not only the fossils on display, but the people who unearthed them. This unique tour was developed specially to celebrate National Fossil Day.
- The Illinois State Geological Survey shared toolkits and geoscience materials to celebrate Earth Science Week.
- A finalist in the Earth Science Week photography contest included Aaliyah Craven of Lemont, Illinois.

Indiana
- The Indiana Geological Survey promoted Earth Science Week among students, teachers, and others across the state by sharing toolkits and geoscience materials.
- For National Fossil Day, the Children's Museum of Indianapolis held Paleopalooza, a fun-filled event with activities such as a hands-on dinosaur station, a *T. rex* photo display, a special performance from their interactive dino show, and much more.
- Falls of the Ohio Foundation invited guests to celebrate National Fossil Day at the Falls of the Ohio State Park with free admission.

Iowa
- The University of Northern Iowa and BMC Aggregates partnered for the "Sunday at the Quarry" event in Waterloo to offer the public opportunities to dig for fossils, search for geodes, and interact with Earth systems.

Kansas
- The Kansas Geological Survey obtained and distributed Earth Science Week Toolkits for educators.
- National Fossil Day at the Sternberg Museum of Natural History invited

guests to join them as they explored this year's theme "The Age of Reptiles: Not Just Dinosaurs".
- Fort Hays State University's Department of Geosciences hosted events and contests throughout the month of October in celebration of Earth Science Week.

Kentucky
- The Kentucky Geological Survey hosted its annual Open House to celebrate Earth Science Week. Attendees were given free kits and were invited to browse the rock and fossil collections under microscopes, observe demonstrations, and more.

Louisiana
- Poverty Point World Heritage Site in Pioneer hosted activities for International Archeology Day, including a guided tram tour with the station archaeologist, a demonstration of prehistoric tools, and more.

Maine
- The Main Geological Survey promoted Earth Science Week and distributed kits among educators statewide.
- Visitors had fun exploring fossils from the L.C. Bates Museum collection, doing a scavenger hunt, making fossil molds, learning about plant, animal and track fossils and more.

Maryland
- The Maryland Geological Survey distributed kits among educators in celebration of Earth Science Week.
- Children ages 4–10 were invited to be paleontologists for the day at the Calvert Marine Museum. Ongoing activities included digging for "dinosaur bones" in the Discovery Room sandbox, a scavenger hunt filled with fun dinosaur facts, and making a dinosaur hat.

Massachusetts
- The Oak Bluffs Library held a National Fossil Day event for Earth Science Week. People were invited to and share their fossils with collectors and listen to presentations.
- The Massachusetts Geological Survey promoted Earth Science Week.
- In celebration of Earth Science Week, the Harvard Museum of Natural History welcomed visitors to learn about current research in Earth science through hands-on activities led by graduate students in Harvard's Department of Earth and Planetary Sciences.

Michigan
- The Michigan Geological Survey promoted Earth Science Week and distributed kits among educators statewide.
- Fernwood Botanical Garden and Nature Preserve celebrated National Fossil Day with an amazing display of fossils from Fernwood's collection.
- The Michigan State University Museum, in partnership with Michigan Earth Science Teachers Association, celebrated National Fossil Day on Sunday, October 14 with an afternoon of fossil fun. Activities included a fossil dig and sift, a Junior Paleontologist hunt, fossil augmented reality, and much more.
- Michigan Technological University hosted an Energy & Environment Day event in September as part of its CareerFEST. Students had opportunities to learn about different job and career opportunities.

Minnesota
- Winona State University's National Fossil Day exhibit drew a steady crowd of students, faculty, and interested community members.
- The Science Museum of Minnesota celebrated National Fossil Day on Saturday, October 13th

with various extra fossil-related stations set up throughout the museum.

- The National Association of Geoscience Teachers promoted Earth Science Week among teachers and citizens.

Mississippi

- A "Fossil Extravaganza" open house and a "Fossil Art & Story Contest" were sponsored by Dunn-Seiler Museum and the Department of Geosciences at Mississippi State University with tours, family activities, free fossils, and refreshments.

Missouri

- The governor of Missouri recognized October 14–20 as Earth Science Week 2018.
- The Missouri Geological Survey promoted Earth Science Week among teachers and citizens with the distribution of educational materials.
- The University of Missouri Museum of Art and Archaeology celebrated International Archaeology Day, disseminating information on archaeology and related activities statewide.
- The Bollinger County Museum of Natural History held a National Fossil Day celebration in Marble Hill, Missouri on Saturday, October 14 from 12:00 (noon) to 4:00 p.m.
- The Missouri Department of Natural Resources promoted Earth Science Week.
- The Missouri State Museum was a great place to visit throughout Earth Science Week, and especially on National Fossil Day, to see fossils in the limestone walls, floors and stairs of the State Capitol.
- Visitors to the Ed Clark Museum of Missouri Geology received a small Crinoid fossil (the Missouri state fossil) in celebration of Earth Science Week.

Montana

- Montana State University's Museum of the Rockies invited children to earn "Junior Paleontologist" badges by participating in National Fossil Day activities.
- The University of Montana celebrated National Fossil Day at the UM Paleontology Center (UMPC) for an open house! The event featured tours of the displays and the research collections, fossil identifications, kids' activities and more.
- Carter County Museum provided paleo-themed activities include "making your own amber," microsite identification and sorting, fossil cast painting and more during museum hours.
- The inaugural National Fossil Day event was held at Makoshika State Park.
- The Museum of the Rockies (MOR), the Children's Museum of Bozeman (CMB), and the Bozeman Public Library (BPL) celebrated National Fossil Day with dinosaur-themed activities all day long at all three locations.

Nebraska

- Agate Fossil Beds National Monument celebrated National Fossil Day with special events.
- The Nebraska Conservation and Survey Division distributed Earth Science Week kits among teachers.
- University of Nebraska State Museum invited children and families to celebrate National Fossil Day at their special paleontology themed night at Morrill Hall. The public had opportunities to observe paleontologists doing fossil prep, explore hands-on fossil "digs", and more.

Nevada

- The governor of Nevada issued a perpetual proclamation of Earth Science Week.

- The Nevada Bureau of Mines and Geology sponsored a field trip, "Sparkling or Still? A Tour of the Geology from Soda Lakes to Stillwater Marsh, Nevada," in celebration of Earth Science Week.
- Guests joined the Protectors of Tule Springs and the Las Vegas Natural History Museum National Fossil Day as they celebrated fossil discovery in Southern Nevada.
- The Nevada Bureau of Mines and Geology also offered a scavenger hunt promoting participants to explore a unique part of the cultural and geologic history of Carson City at various places around the city including the Stewart Indian School.
- The Nevada Bureau of Mines and Geology distributed Earth Science Week Toolkits and Geologic Map Day materials to teachers.

New Hampshire

- The New Hampshire Geological Survey promoted awareness of Earth Science Week among residents and community members.

New Jersey

- The New Jersey Geological and Water Survey distributed Earth Science Week Toolkits and Geologic Map Day materials to teachers.
- The governor of New Jersey issued a proclamation recognizing Earth Science Week 2018.
- The New Jersey State Museum invited the public to join them in celebrating International Archaeology Day 2018 by exploring New Jersey's wonderful heritage of fossils, on display throughout the Natural History Hall, and by asking museum scientists any questions you may have about New Jersey natural history.
- Finalists in the Earth Science Week visual arts contest included Janiru Sumanasiri of Nutley, New Jersey.

New Mexico

- The New Mexico Museum of Natural History & Science hosted a National Fossil Day event featuring hands-on activities, special fossil displays, and giveaways. Kids were also able to be sworn in as a Junior Paleontologist through the National Park Service during a special ceremony.
- Mesalands Dinosaur Museum and Natural Science Laboratory invited visitors to attend their National Fossil Day event.

New York

- Earth Science Week 2018 Toolkits were distributed to students, teachers, and others by representatives of the Geology Research Center of the New York State Museum.
- Finalists in this years Earth Science Week essay contest include Jaeho Lee of Rexford, NY.
- The American Museum of Natural History held a National Fossil Day inspired event to help students understand how studying fossils teaches us about the history of life, past climates, and ancient landscapes. The day consisted of activities such as the exploration of fossil formation, extinction theories, fossil diversity and much more.
- The Museum of the Earth and the Paleontological Research Institution hosted Fossil Mania, a three-hour long event that educated museum visitors about fossil types, & formation with hands-on specimen driven activities.
- To coincide with Earth Science Week, the GeoMentors (State University of New York at Fredonia YouthMappers Chapter) hosted a GeoMentors Map-A-Thon event at McEwen Hall on their campus.

North Carolina

- The North Carolina Geological Survey promoted awareness of Earth Science Week among residents and community members.

- Namit Vernekar of Charlotte, North Carolina, won first place in the visual arts contest with a creative and colorful drawing of two sides (real and imaginary) of Earth.
- The Aurora Fossil Museum celebrated National Fossil Day with a variety of special guests, live music, events and activities planned for families and kids. The free event is held annually through a partnership with the National Park Service.
- Imagination Station Science and History Museum hosted a National Fossil Day inspired Educator Trek – Fossil Hunt through the waters of Green Mill Run.
- Saachi Tamboli and Abhiraj Das were also recognized as finalists in Earth Science Week's visual arts contest.
- Earth Science Week 2018 Toolkits were distributed to students, teachers, and others by representatives of the History of Earth Sciences Society.

North Dakota

- North Dakota's governor issued a perpetual proclamation of Earth Science Week.
- North Dakota State University's (NDSU) Herpetology Club, NDSU's Geosciences Department, and the Fargo Public Library invited the public to join them in celebration of National Fossil Day.

Ohio

- The Ohio Department of Natural Resources (ODNR) Division of Geological Survey distributed Earth Science Week Toolkits and other materials to schools and others.
- The Ohio Division of Geological Survey celebrated Earth Science Week by leading the public on several geologic hikes and educational events throughout the state.

- Cincinnati Museum Center encouraged the public to participate in National Fossil Day.
- The Ohio Earth Science Teachers Association promoted Earth Science Week among teachers and citizens.
- In celebration of National Fossil Day, Ohio State University and the Ohio Geological Survey ran Ohio Statehouse Fossil Tour, a geologic tour of the fossil-rich building stones of Capitol Square in Columbus.
- The Cleveland Museum of Natural History hosted a full day celebration of Earth Science Week and International Archaeology Day where visitors dug into archaeology, geology, mineralogy and more.

Oklahoma

- Oklahoma's governor issued a perpetual proclamation of Earth Science Week.
- The Oklahoma Geological Survey distributed Earth Science Week Toolkits to educators and students.
- Sam Noble Museum and the Oklahoma Archeological Survey cosponsored an archaeology event in honor of the 25th anniversary of the discovery of the painted bison skull at the Cooper site in Harper County, Oklahoma as well as International Archaeology Day.

Oregon

- The Cannon Beach History Center and Museum partnered with the Archaeology Institute of America for International Archaeology Day. To celebrate, the museum hosted a presentation on Dr. Cameron Smith, human history, archaeology, and evolution in the Americas.
- A fossil-focused National Fossil Day Walk & Talk was hosted by the Museum of Natural and Cultural History.

- The Oregon Department of Geology and Mineral Industries promoted their new interactive geologic map of Oregon for Geologic Map Day. The application allows the public to view the geologic story of Oregon and access supplementary information with a click of a mouse.

Pennsylvania
- The State Museum of Pennsylvania hosted their "Giants of the Late Cretaceous." celebration for National Fossil Day.
- Earth Science Week Toolkits were distributed among science teachers statewide by the Pennsylvania Geological Survey.
- The governor of Pennsylvania declared October 14–20 to be Earth Science Week.
- Earth Science Week 2018 Toolkits were distributed to students, teachers, and others by representatives of the Topographic and Geologic Survey.
- The Newlin Grist Mill invited the public to participate in an archaeology dig to celebrate International Archeology Day.
- On National Fossil Day, Art Wegweiser, PhD, and Professor Emeritus, exhibited crystals and fossils from his personal collection.
- The Pennsylvania Earth Science Teachers Association promoted Earth Science Week among teachers and citizens.
- Penn State College of Earth and Mineral Sciences promoted Earth Science Week on their website.

Puerto Rico
- An International Archaeology Day celebration held at the Castillo de San Cristobal was sponsored by the San Juan National Historic Site, National Park Service, Research Initiatives and Undergraduate Creative Activity, and the Department of Sociology and Anthropology at the University of Puerto Rico, Rio Piedras.

- Ciencia Puerto Rico promoted the celebration of Earth Science Week to their general public.
- Iniciativas de Investigación y Actividad Creativa Subgraduadas (iINAS), Universidad de Puerto Rico-Recinto de Río Piedras, and San Juan National Historic Site, National Park Services (NPS) sponsored a series of symposia for professional archaeologists as well as undergraduate students at the university campus and the NPS facilities. The week consisted of workshops, poster sessions, and ended with student presentations.
- The Symposium of Archeology and Cultural Heritage took place from October 17th until the 19th and included conferences, workshops, guided visits, a student symposium, and a poster session. This event was sponsored by the San Juan National Historic Site, National Park Service, Research Initiatives and Undergraduate Creative Activity, and the Department of Sociology and Anthropology at the University of Puerto Rico, Rio Piedras.

Rhode Island
- The Roger Williams Park Museum of Natural History and Planetarium held its "Fossil Frenzy Weekend" to celebrate National Fossil Day. Visitors viewed fossils from the museum's vaults, went on a fossil quest, did a fossil puzzle, and made their own trilobite masks and more.

South Carolina
- The South Carolina Geological Survey celebrated Earth Science Week by distributing Earth Science Week Toolkits to educators.
- 2,000 Year History Park Working Group, in collaboration with the city of Cayce and the River Alliance, sponsored a series of guided walking tours in which participants learned how people have

used the area's natural resources through the years.
- The Aiken County Historical Museum hosted their first ever Archaeology Day celebration with hands-on interactives and a special viewing of the SRARP film, "Reconstructing Hawthorne.". Organizations helping them celebrate archaeological discoveries in Aiken County included Savannah River Archaeological Research Program (SRARP) and The Hitchcock Woods Foundation.
- Environmental Education in South Carolina promoted Earth Science Week among teachers and citizens.

South Dakota
- South Dakota's governor issued a perpetual proclamation of Earth Science Week.
- South Dakota Discovery Center invited the public to share their favorite Earth science themed photo on social media tagging it #EarthScience.

Tennessee
- The governor of Tennessee declared October 14–20 to be Earth Science Week.
- The Tennessee Department of Environment and Conservation's Tennessee Geological Survey distributed Earth Science Week Toolkits to educators.
- The University of Tennessee's McClung Museum of Natural History and Culture with support from the Archeological Institute of America hosted the "Can You Dig It?" event to celebrate International Archeology Day and National Fossil Day.

Texas
- In partnership with the Austin Earth Science Week Consortium, the Bureau of Economic Geology held a career event for middle school students. Earth science

professionals gave presentations about their careers, and students were able to participate in hands-on activities through various exhibits. Students heard from local STEM (science, technology, engineering, and math) professionals including geologists, geophysicists, engineers, hydrologists, meteorologists, paleontologists, water conservation specialists, biologists, and aerospace engineers.

- *Discover Earth Science!*, a free outdoor learning event sponsored by Texas Master Naturalist, Rio Brazos Chapter and Acton Nature Center featured geoscience-focused exhibits; demonstrations; special presentations; activities for children; nature walks featuring geology, soils, landscape formation, and fossils; and Earth-inspired art, music, and literature.
- The Dallas Paleontological Society hosted a National Fossil Day celebration at the Heard Natural Science Museum & Wildlife Sanctuary.
- The Texas Memorial Museum held a public event, in celebration of National Fossil Day, with activities including fossil identifications, a paleontologist meet and greet, and more.
- Locals, students, visiting scientists, fossil hunters and enthusiasts gathered in Fannin County for a week of activities and events to partake in Earth Science Week.
- The Wiess Energy Hall at the Houston Museum of Natural Science hosted an Earth Science Week field trip throughout the building to journey from the Big Bang to the Houston of the future.
- Baylor University's Mayborn Museum invited people to explore fossils by digging through gravel, searching for fossils, making their own fossil, and examining rising sea levels during their event.
- Waco Mammoth National Monument invited the public to attend the City of Waco's Fall Fossil Festival in celebration of National Fossil Day.

- Organizations including the local Bois d' Arc Chapter of Texas Master Naturalist teamed with the Dallas Paleontological Society, the Ladonia Chamber of Commerce, the City of Ladonia, the Blackland Prairie Chapter of Texas Master Naturalist, and the Ladonia Volunteer Fire Department contributed to the 2018 celebration of Earth Science Week.
- The Houston Geological Society held various events for Earth Science Week with partners in the Houston area, including field trips and a special celebration at the Houston Museum of Natural Science.
- The Houston Energy Day Festival, sponsored by around 100 businesses in Houston, brought over 25,000 people out to view demonstrations, learn about energy science, and play games to compete for prizes and giveaways.
- The Just Energy Foundation awarded top-performing students of the Science and Engineering Fair of Houston during the Energy Day Festival.
- Consumer Energy Alliance partnered with the Houston Museum of Natural Science Wiess Energy Hall's Energy & Conservation Club to host the eighth annual Art, Essay, and Media contest in Houston. Winning students and their teachers received awards and were recognized on stage at Energy Day.
- The Art Center of Waco featured an Art Expedition at Waco Mammoth National Monument during the Fall Fossil Festival.
- Texas Memorial Museum celebrated National Fossil Day in Austin, where the public was invited to meet paleontologists and learn about the featured fossil finds.
- The Geology Department at Texas A&M University Kingsville kicked off a week of celebration with an open house, several activities

including GIS, paleontology, mineralogy and petrology, field geography and geophysics demonstrations. Personal collections of the faculty, including meteorites, minerals and fossils were also on display.

- Earth Science Week 2018 Toolkits were distributed to students, teachers, and others by representatives of the International Medical Geology Association and the Petroleum History Institute.
- The National Park Service recognized the City of Ladonia in Fannin County, Texas as a sponsoring participant of National Fossil Day. The Dallas Paleontological Society, and the Ladonia Fossil Park held additional Earth Science Week activities.

Utah

- Glen Canyon National Recreation Area hosted a celebration of National Fossil Day where visitors could go on a fossil walk, have fun with fossils, ask a scientist questions, and more.
- Dinosaur National Monument celebrated National Fossil Day with events throughout Earth Science Week.
- The Utah Geological Society distributed Earth Science Week Toolkits to educators and students.
- The 3rd annual Moab Festival of Science was sponsored by organizations such as the US Geological Survey, Utah Department of Natural Resources, the National Park Service, the Department of Energy, the Museum of Moab, and many others. This event served to connect and inspire the citizens of southeastern Utah with the wonders of science and the thrill of scientific discovery with its numerous hands-on activities and awe-inspiring tours.
- Earth Science Week 2018 Toolkits were distributed to students, teachers, and others by representatives of the Utah Geological Survey.

Vermont

- The Vermont Geological Survey shared geoscience materials to celebrate Earth Science Week.

Virginia

- In celebration of International Archaeology Day, the Virginia Museum of Natural History held a fun afternoon event for visitors to learn about the museum's fascinating archaeology research and collections while getting a glimpse at what goes on behind-the-scenes in the labs and collections areas.
- The Torpedo Factory Art Center, in partnership with the American Geosciences Institute, invited the public to attend an exciting afternoon of music, interactive art, and more at their 23rd annual Art Safari event.
- Tri-County/City Soil & Water Conservation District hosted an event at the Spotsylvania Town Center Library in the Mall to celebrate Earth Science Week. This celebration featured fun games and activities for all ages and invited guests to find their inspiration in rocks, soil, and water.
- The Virginia Department of Mines, Minerals and Energy distributed Earth Science Week Toolkits to teachers in the state.
- Esther Gammill of Virginia Beach, Virginia was selected as a finalist of Earth Science Week's visual arts contest with her piece depicting how art is poured into the Earth.
- In honor of Earth Science Week, the James Madison University (JMU) Geology Club and Ecology Club held a guided nature walk through the Edith J. Carrier Arboretum. JMU students led community members of all ages on a journey to explore the Arboretum's natural history, learn about Virginia's common trees and other foliage, rock formations, and soils under the autumn leaves!
- Caden Longwater, Lluvia Perez, and Vikram Kolli were recognized as finalists the Earth Science Week essay contest with their papers addressing this year's theme, "Finding 'Art' in Earth."
- The National Science Foundation featured an Earth Science Week edition of their Four Awesome Discoveries video series.
- The National Park Service hosted a belated celebration of Earth Science Week at the Historic Jamestown Visitor Center. There they had educational demonstrations on the Earth processes that help inspire the natural and cultural history of Colonial National Historical Park.
- The U.S. Geological Survey shared "Revealing Our Planet Through Earth As Art," satellite images that brilliantly combined science and art to showcase the beauty and wonder of our planet. This series aligned perfectly with the 2018 Earth Science Week theme "Earth as Inspiration."
- Earth Science Week 2018 Toolkits were distributed to students, teachers, and others by representatives of the National Center for Earth and Environmental Nanotechnology.
- The Virginia Association of Science Teachers promoted Earth Science Week among teachers and citizens.

Washington

- The Burke Museum of Natural History and Culture celebrated National Fossil Day with several activities for the public including live fossil preparation. The public was also invited to learn about the ongoing excavation of a Tyrannosaurus rex skeleton.
- In recognition of Archaeology Month, Grant PUD and the Wanapum invited members of the public to Archaeology Days 2018. Archaeology Days provided a fun and interactive experience for all to learn more about the Wanapum way of life. Those attending heard from speakers, saw demonstrations and participated in various activities.
- White Pass Country Historical Museum hosted a presentation by archaeologist Rick McClure who provided an overview of the history and archaeology of the Cowlitz-Yakama Trail.
- In honor of International Archaeology Day, Cowlitz County Historical Museum featured a presentation on the environmental setting, site stratigraphy and content, and age, of the known locations for Washington's earliest settlers.
- In cooperation with the Departments of Classics at the University of Washington and the University of Puget Sound, the Puget Sound Society co-sponsored an annual lecture series that introduced audiences to the latest archaeological research and discoveries.
- Earth Science Week Toolkits were distributed among science teachers statewide by the Washington Geological Survey and the Washington State Department of Natural Resources.

West Virginia

- Grave Creek Mound Archaeological Complex hosted an International Archaeology Day event with a behind-the-scenes tour of the research area, spear throwing, pottery making, a museum hunt, exhibits and demonstrations relating to archaeology, and more.

Wisconsin

- Earth Science Week 2018 Toolkits were distributed to students, teachers, and others by representatives of the Wisconsin Geological & Natural History Survey as well as the Soil Science Society of America.
- Matt Meisenheimer of Janesville, Wisconsin, won first place in the photo contest with a captivating image of Kalalau Valley in Kauai, Hawaii.

Wyoming

- The Wyoming State Geological Survey disseminated Earth Science Week 2018 Toolkits among students and educators statewide.
- Fossil Butte National Monument staff and volunteers held a celebration of National Fossil Day with various activities, including hands-on demonstrations, arts and crafts and exhibit tours of the monument museum.
- Tate Geological Museum at Casper College hosted a National Fossil Day Open House with fun fossil activities, crafts, tours and treats.
- The Wyoming State Geological Survey and University of Wyoming Geological Museum celebrated with a geoscience event on October 13th.
- The Buffalo Bill Center of the West in Cody held a field trip for local elementary school students on International Archeology Day. Students learned differences between paleontology and archeology.

International Events

According to Google Analytics, the Earth Science Week website was accessed by users in 212 countries, territories, and regions in 2018. More substantial activity included:

Australia

- Geoscience Australia's Earth Science Week 2018 celebrated the theme of "Earth as Inspiration."
- Australians celebrated National Fossil Day by touring the world class collection of amazing specimens stored in the National Fossil and Mineral Collection.
- Australia's premier science festival capped months of productive work across the state by staff promoting geological resources. It was also a suitable time to feature several new maps highlighting the diversity of work done by the Geological Survey of New South Wales (GSNSW).
- Thanks to the GSNSW, two free events are open to the public in honor of Earth Science Week. A free public tour along Newcastle's rocky coastline led by expert geoscientists as well as a presentation on the Earth's inner workings, and how shifting plate tectonics shape our world and long-term climate change.
- The GSNSW also engaged the community throughout the week with science related events and resources.

Belize

- The National Institute of Culture and History (NICH) in partnership with the Corozal House of Culture, and The Ministry of Education hosted their annual International Archaeology Day in Coroza. Some of the exhibits, which were open to the public, included: Maya crafts for rituals, utensils used daily, obsidian, jade, cherts and flints, grinding stone, stone tools, textiles, and much more.

Brazil

- To celebrate International Archeology Day 2018, the Olho D'Água Institute organized the workshop "Arte na Serra" in the living area of the Atelier and Dona Graça Library.

Canada

- The Oshawa Museum in Ontario partnered with Trent University Durham to host an International Archeology Day event featuring interactive displays, engaging activities, and lectures on archeology.
- The 4th International Olympiad of Geography was held in Quebec City, Canada engaging 43 participating countries.
- The Royal Ontario Museum celebrated International Archaeology Day with a free day of activities and exhibits.

China

- Teams of educators and students entered Earth Science Week's Earth Expressions video contest. Finalists included a group of 15 students from Nanjing Hankai Academy in China.

Colombia

- Events were held throughout the week in the Bucaramanga Metropolitan Area. These lessons were led by geology students at the Industrial University of Santander in Santander, Colombia. Participants included students and university staff from Piedecuesta and Bucaramanga.
- Earth Science Week-related education materials (in English and Spanish) were distributed.

Czech Republic

- Museum of the Bohemian Forest in Tachov and Charles University in Prague sponsored "Metals in the Past", an event for school children to learn more about the ways of production and use of metals in human history.

- The Archeoklub Kadaň and the Municipal Museum in Kadaň offered a bus trip to the archeological park in Březno by Louny.
- International Archaeology Day was celebrated for the first time in Liliová thanks to participation from the National Heritage Institute, Directorate-General.
- Muzeum T. G. M. Rakovník sponsored an educational walk commemorating the 100th anniversary of the founding of the Czechoslovak Republic.
- The Archaeological department of Czech National Museum offered to the public a rich program full of activities and information about archaeology, archaeologists, their work and results.
- The City of Prague Museum celebrated International Archeology Day by hosting an event to promote awareness of archeologists' work.

Georgia
- The University of Georgia's departments of Archaeology, Anthropology and Art held their annual educational event in the village Samshvilde. The event included various workshops as well as a lecture.

Greece
- The Ephorate of Antiquities of Achaea, Ministry of Culture and Sports, Patras, Greece celebrated this year's International Archaeology Day with a two-session educational event that was held on October 18th and 19th, 2018 at the Mycenaean Archaeological Park of Voudeni, Patras.

Guyana
- The University of Guyana celebrated archaeology and the thrill of discovery with a movie and facilitated discussion as

well as an educational and interactive exhibition.

India
- In Shukrawar Peth, Pune, Maharashtra a three-month long certification course on Earth science was organized by Jeevidha, an NGO working in environment awareness and conservation, featuring 24 classroom sessions on Fridays & Saturdays plus seven field visits on Sundays.
- In observation of Earth Science Week and it's 2018 theme of "Earth as Inspiration", Jividha organized three events including a certificate course in Earth Science, an exhibition on 'Man & Stone', as well as a study tour of Grater Rann of Kutch.
- An International Archaeology Day Ceremony was sponsored by the Subharti School of Buddhist Studies, Swami Vivekanand Subharti University, Meerut, U.P, India.

Iran
- The Department of Haoma organized an event at the Neshat Primary School on International Archaeological Day. The goal was to introduce children to the village period in Iran, as well as acquaintance with the process of exploring an ancient hill. Through the story, children became familiar with how people lived, architectural spaces, livelihoods in rural areas, and eventually the formation of an ancient site.
- Horizon of Archaeology, a session about Archaeology and its sub-branches, was sponsored by Society of Iranian Archaeologists.

Ireland
- Earth Science Week events took place in museums, libraries, schools, universities, geoparks and more, including urban geowalks, family activity days, talks, conservation days and fieldtrips.

Italy
- A Brief History of Millennial Knowledge was sponsored by Parco Archeologico di Travo (Pc), Italy. This event featured experimental demonstrations and workshops for children.

Japan
- The first officially Earth Science Week celebration in Japan was held in Shizuoka from October 13 through 19 mainly at Museum of Natural and Environment History, Shizuoka. More than 1,800 people participated in this inaugural series of events.

Macedonia
- To celebrate International Archaeology Day 2018, the Center for Scientific Research and Promotion of Culture "HAEMUS" organized a series of lectures in the field of archaeology starting at the EU Infocentre in Skopje.

Netherlands
- As a precursor to Archeology Days, and a kickoff for the Excavated History Month in the European Year of Cultural Heritage, the Stichting Nationale Archeologiedagen / National Archeology Days Foundation organized the Archeonacht in the Rijksmuseum van Oudheden.
- Stichting Nationale Archeologiedagen / National Archeology Days Foundation; and numerous others throughout the Netherlands hosted various celebrations of International Archaeology Day.

Peru
- Peru's Nivín Archaeology Museum hosted a week of activities for Earth Science Week's International Archaeology Day.

Russia

- Along with more than 100 Collaborating Organizations, the Archeological Institute of America promoted celebrations of International Archeology Day in cities around the globe including Samara, Russia.

Saint Lucia

- The National Emergency Management Organization joined the nation of Saint Lucia in celebration of Earth Science Week with a series of activities planned around the island to commemorate the week. These activities encompassed the participation of schools, communities, the public and private sectors; and were facilitated by a team from the UWI Seismic Research Centre from Trinidad and Tobago.

Thailand

- The 2nd International Olympiad of Earth Sciences was held at the Mahidol University Kanchanaburi Campus in Kanchanaburi, Thailand.

Trinidad & Tobago

- Under a collaborative effort by the National Emergency Management Organization (NEMO), the University of the West Indies Seismic Research Centre and other area organizations, experts offered workshops, exhibits and seminars on geoscience in celebration of Earth Science Week.

United Kingdom

- The Geological Society of London (GSL) supported several Earth Science Week events including geowalks and a school workshop.
- UK Earth Science Week and GSL sponsored geowalks, school visits, and more.
- GSL also launched a photography competition for Earth Science Week with the theme "Earth science in our lives." Winning photos are included in GSL's 2019 calendar.

See Our News Coverage

Because of the large and increasing number of news clippings citing Earth Science Week activities and resources, the print edition of this report no longer includes clippings. To view the hundreds of press releases and news items promoting awareness of Earth Science Week each year, please visit online at **www.earthsciweek.org/highlights**. Thank you for helping us in our efforts to conserve resources and protect the environment.

Earth Science Week 2018 Visual Arts Contest entry by Esther Gammill.

www.ingramcontent.com/pod-product-compliance
Lightning Source LLC
Chambersburg PA
CBHW041303180526
45172CB00003B/945